T0342445

PRAISE FOR *WHEN THE HOOD COMES OFF*

"A fascinating look at race and racism online. Drawing on sensitive interviews with those who must grapple with racist discourse, complemented by analyses of large-scale patterns in social media, Rob Eschmann digs into the depths of how technologies for social interaction have shaped our nation's continuous struggle with race. Informed by the author's personal experiences with racist discourse, *When the Hood Comes Off* is a consistently sensitive and probing book."

—MARIO SMALL, author of *Qualitative Literacy: A Guide to Evaluating Ethnographic and Interview Research*

"While many great works ask us to diagnose racism as it has happened, *When the Hood Comes Off* stands to change the way we think about the racism ahead of us, unfolding in digital space."

—SAIDA GRUNDY, author of *Respectable: Politics and Paradox in Making the Morehouse Man*

"Eschmann documents how the subtle racism that people of Color experience in public spaces becomes crass, aggressive, and direct in private, online settings. Silent or hesitant on race topics when in public, far too many Whites become almost Klan-like when interacting with people of Color in the anonymous online sphere. *When the Hood Comes Off* is splashed with memorable, poignant personal examples and written in a highly readable and engaged way."

—EDUARDO BONILLA-SILVA, James B. Duke Distinguished Professor of Sociology, Duke University, and author of *Racism without Racists: Color-Blind Racism and the Persistence of Racial Inequality in America*

"Eschmann raises important issues about what it means to be a young person in our very online world at a time when White supremacy is resurgent. He draws on a huge trove of data and writes about it with lucidity and urgency."

—JESSIE DANIELS, author of *Nice White Ladies: The Truth about White Supremacy, Our Role in It, and How We Can Help Dismantle It*

"Eschmann does not limit himself solely to studying what happens on social media platforms. He interviews students of Color all across the country who experience online racist rhetoric and who forge online and offline spaces of antiracist resistance."

—IBRAM X. KENDI, author of *How to Be an Antiracist* and *Stamped from the Beginning: The Definitive History of Racist Ideas in America*

WHEN THE HOOD COMES OFF

The publisher and the University of California Press Foundation gratefully acknowledge the generous support of the Lawrence Grauman, Jr. Fund.

WHEN THE HOOD COMES OFF

Racism and Resistance in the Digital Age

ROB ESCHMANN

UNIVERSITY OF CALIFORNIA PRESS

University of California Press
Oakland, California

© 2023 by Robert Eschmann

Library of Congress Cataloging-in-Publication Data

Names: Eschmann, Rob, author.
Title: When the hood comes off : racism and resistance in the digital age /
    Rob Eschmann.
Description: Oakland, California : University of California Press, [2023] |
    Includes bibliographical references and index.
Identifiers: LCCN 2022052938 (print) | LCCN 2022052939 (ebook) |
    ISBN 9780520379725 (cloth) | ISBN 9780520379749 (paperback) |
    ISBN 9780520976894 (ebook)
Subjects: LCSH: Racism in mass media. | Digital communications—Social
    aspects—United States. | Online social networks—United States. |
    Anti-racism--United States. | African Americans in mass media. |
    Minorities in mass media.
Classification: LCC P94.5.M552 U629 2023 (print) | LCC P94.5.M552 (ebook) |
    DDC 302.23089—dc23/eng/20221230
LC record available at https://lccn.loc.gov/2022052938
LC ebook record available at https://lccn.loc.gov/2022052939

32 31 30 29 28 27 26 25 24 23
10 9 8 7 6 5 4 3 2 1

For Malachi, Karis, and JD

Contents

1

AN INTELLECTUAL PUZZLE

THE FIRST TIME I played video games on the internet was also the first time I was called the n-word.

During my freshman year of college, I spent an inordinate amount of time playing video games. For hours each day, my friends and I took turns fighting aliens (or one another) in a digital world, with occasional breaks for studying. While today any smartphone can play video games over the internet, back then it was more complex. Our campus network wouldn't allow video-game consoles to connect to the internet, but we figured out how to play on the local network or with other students on campus. Great rivalries were formed between dorms as we challenged one another in *Halo,* a popular combat-based game, the winners gaining bragging rights across campus and the losers sometimes resorting to physical pranks as payback.

Over winter break, I went to visit two cousins around my age, CJ and David, who lived a few hours away by train. Not only did they play video games even more than I did, but they also had access to playing online, something I was eager to try. The ability to play and communicate with people who could be anywhere around the world was alluring.

Apart from the competition being a bit stronger, playing *Halo* online didn't feel much different than playing on campus. We weren't using the headset to chat with the people we were playing against, so it was just us, goofing around while playing, as usual. But I thought that chatting was the main draw of playing online, and I wanted to get the full online gaming experience. Wasn't making friends from around the globe the point?

When I asked my cousins if we could plug the headset in, they looked at each other uneasily and said, "No, we don't use the headset." When I asked why, they replied, "Because every time we do, we get called the n-word."

This was hard for me to believe. As a Black man growing up in Chicago and now living on a predominantly White campus, I knew that racism was real. I was accustomed to being stopped by the police for no reason and had experienced countless racial microaggressions—subtle slights that question the intelligence, appearance, or lawfulness of people of Color.[1] Beyond my personal experiences, I had taken sociology classes that taught me to think critically about racial inequality in education, mass incarceration, and other forms of racial oppression. But I had *never* been called the n-word maliciously. It's just not something that happened in my neighborhood growing up, which was majority Black and Latinx, or at school.

So I didn't believe them. "You mean it's happened before," I said, "not that it happens every time."

"No," they said. "We mean every single time."

Still incredulous, I had to try it for myself. So, I put on the headset and simply said, "Hey, what's up guys," into the microphone.

The response was something I'll never forget. I was playing with an online username, GalacticHair. And the first words in response to my greeting were, "Your username should be GalacticNigger."

I couldn't believe it. For the rest of the game, the guys we were playing continued throwing hate speech at me. I didn't back down and spent the game focusing more on the war of words than the virtual combat. When our team lost badly, one of my cousins pointed out that this was the reason they didn't like to chat while playing. "You get so mad that you forget how to play; then you end up losing to racists."

It was like losing twice.

Even though my cousins "told me so," I still could not believe that this was a regular occurrence. I insisted that we must have run into some bad apples by chance—maybe those guys were in the KKK—but there's no way that everyone online talks that way. So I kept the headset on for the next four or five games. Each time we matched against different users but experienced the same result.

At some point I just decided to take the headset off. My cousins were right: to have fun playing this game without being harassed, we couldn't engage in the chat function.

When winter break ended and I went back to college, I wondered how many of my White friends might have been also playing *Halo* online over winter break. And I thought, if *everyone* we encountered online used racist language, is that the type of behavior my friends are engaging in when I'm not around? Might some of them have been behind the usernames and microphones that were hurling racial slurs at my cousins and me?

This question was the beginning of an intellectual puzzle: what does the style or expression of racist language in a virtual world, or on the internet, tell us about the nature of racism in the real world?

A few months later, I was hanging in a friend's room on campus where of course the Xbox was on, and we were taking turns playing

Halo. There were four or five guys in the room, all of them White except for me. Music was playing, much of it acoustic songs that I didn't recognize. But then, to my surprise, Tupac Shakur's *Changes* came on. Everyone in the room, including myself, rapped along with Pac, singing:

> I see no changes
> Wake up in the morning and I ask myself
> Is life worth living should I blast myself
> I'm tired of being poor, and even worse, I'm Black
> My stomach hurts, so I'm lookin' for a purse to snatch

Growing up, my mother didn't allow my sister and me to play many explicit rap songs at home. But Momma had a few favorites that played on repeat: Lauryn Hill's *Miseducation* album, and two songs from Tupac's *Greatest Hits,* "Changes," which has a few curse words but was a deep exploration of racism and poverty, and "Dear Mama," another meaningful song that had none of Pac's violent or misogynistic lyrics. Knowing how different our tastes in music were, I thought it was cool that these guys knew a song that was so near and dear to my heart. But then I remembered what the next lines were, and stopped rapping. I looked around the room at my friends, and wondered, would they sing the n-word along with Pac?

> Cops give a damn about a negro
> Pull the trigger, kill a nigga, he's a hero

No one said it. Everyone stopped rapping that at that lyric and picked back up after the slur. A few of the guys glanced over at me as they did. I laughed out loud and asked for the music to be shut off.

"Yo, I'm not mad," I said, trying to make sure they wouldn't be defensive. "But be honest. If I wasn't here, . . . would you all have rapped *nigga?*"

"No, of course not!"

"No way bro, we would never!"

"C'mon man, how could you think that?"

But one guy in the room, a soccer player with a penchant for being a rebel, had a smirk on his face.

"Oh please," he said, looking around the room. "Yea, you do."

A few of my friends looked sheepish as he continued, saying, "I say it, but not to be racist, just because I like the song."

A few of the guys nodded, admitting that they rapped the n-word from time to time, but assured me they didn't mean anything by it. At least one of them vigorously denied ever doing so, and I believed him. We put the music back on and kept playing video games. I told my good friend Trey, a Black man from Texas who lived one floor up, about the incident later. We had no shortage of race-related stories to tell each other that first year in the dorms.

This experience answered some of the questions I had about how prevalent racial slurs were when playing video games online. Some of my White friends believed that singing the n-word was innocuous, not malicious. But I didn't see a big difference between singing the n-word and shouting it while playing video games. The kids who hurled racial slurs online probably saw that as being pretty innocuous, too. They might not have seen themselves as real racists; they were just using the word as an insult, or to be cool, imitating their favorite rappers.

Following this logic, I reasoned that if my friends were okay saying the n-word as long as no one Black is around to make them feel uncomfortable, they might also be okay saying it while gaming, when they don't have to look a Black person in the eye. Of course, it's possible that none of them ever did! I'm not making an accusation. I'm making a point about how the combination of my online and in-person experiences influenced the way I understood the world around me. I learned that the way White people treated me when we were face to face (overwhelmingly with kindness) did not mean that they didn't do, think, or say racist things in private or when I was not around.

PIECING THE PUZZLE

As a first-year graduate student, I was assigned Eduardo Bonilla-Silva's seminal text, *Racism without Racists*.[2] In the book, Bonilla-Silva discusses what he calls colorblind racism, a racist ideology that sustains unequal racial systems without reverting to hostile language around racial differences. Bonilla-Silva suggests that racist ideologies and racist language are used to legitimate racist social systems. During the Jim Crow era,[3] racist language needed to be explicit in order to justify a harsh and formal racial-stratification system. But contemporary racial systems, which rely less on the explicit and legal separation of races, can be sustained with subtler, milder expressions of racism.

It struck me that while this framework explained the subtle presentations of racist ideologies in many contexts, including most college campuses, it did not explain why overt racism persisted, and even seemed to be the norm, in some online spaces. How could I reconcile my experiences with overt racist language on the internet with Bonilla-Silva's theory? What is it about online spaces that seem to invite a reversion to Jim Crow–era racist language?

Research on the difference between "computer-mediated communication" and face-to-face communication makes it clear that the internet represents a distinct social environment, and that communication in many online contexts may not be bound by the same norms as in face-to-face interaction.[4] For example, Kishonna Gray explores how anonymity and online norms such as "disinhibition" (saying or doing things one might not offline) shape the ways Black and Latinx women experience racism, sexism, and homophobia while playing online video games.[5] Jessie Daniels has written about online White supremacy for decades, from the ways White supremacist organizations use technology to spread their ideologies to (with Matthew Hughey) the proliferation of racist comments on mainstream news sites.[6] The research on online racism adds complexity to our understanding of the ways expressions of racism change in different contexts.

As young people spend more time socializing and learning in online spaces, it follows that their online interactions will have an increasingly large impact on their lives and well-being. Research on racial socialization finds that preparing Black children for experiences with racial bias can lessen the negative impact that stereotypes have on their social and emotional health.[7] But what happens if we socialize children to prepare them to deal with colorblind-style racism, and fail to recognize that they are actually being exposed to overt expressions of racism in online spaces, where adult moderation is less prevalent?

It's difficult to keep kids safe online. I study online abuse for a living, but my kids still surprise me with their online communication experiences (often game-based) on mediums that weren't on my radar. For example, I know how to change settings on video-game consoles so that my kids are able to chat only with their friends (and not with random people online). But I did not know that public chat was a function on the Oculus VR headset that my oldest child, Malachi, had saved his money to buy until his younger sister, Karis, and brother, JD, alerted me to the toxic language people were using on the platform. Malachi assured me he had quickly figured out how to mute the public chat, but this worried me. I had no idea it was going on, nor how to stop it.

I ask some big questions in this book: How does the internet shape the way people talk about race? How are online experiences and responses to racism different from in-person experiences? What impact do these changes have specifically on people of Color, and on society?

My experience with a strange, new, and more explicit style of racism online was like a puzzle piece that didn't fit: how could such explicit racist language be ubiquitous in online spaces, when it was so rare in face-to-face settings? Then, my experience on campus, seeing my friends censor themselves because of my presence, gave me a clue, and I was able to imagine how muted behaviors in person might be connected to the normalcy of more explicit behaviors online. This is what it takes, I think, to fully understand the impact technology has had on the race conundrum: an investigation that pays equal attention to the virtual and physical realities of race and racism, how they differ, and how each influences the other.

This is difficult to do! Often our online realities and behaviors are so different from our day-to-day, face-to-face interactions. In many online spaces, users may not even know the real-world identities of the people they interact with. Online actions are therefore often divorced from perceived consequences in the real world.

The idea that there are separate worlds, one physical and real, and one digital and fake, is known as the "digital dualism fallacy." Social theorist Nathan Jurgenson argues that we actually live in a world that is a mix of the digital and the physical: an augmented reality.[8] There is no better place to study the augmented reality of race and racism than the college campus, a setting where online and in-person communities are less detached, since it is common for students to share classrooms or dining halls with the same peers they engage with on social media. The perceived consequences of online behaviors, therefore, can be more real, or at least more immediate, for these students. If you post something online over the weekend, chances are people in your class on Monday have seen it.

To bridge the gap between the digital and the physical, this book begins on the college campus. I started this eight-year study in 2014 in Chicago at a private, selective university. I conducted interviews with Black, Latinx, and Asian students, as folks of Color are too often denied a voice and are most likely to experience and witness racism and its effects. Centering the perspectives of people of Color is an important part of the critical race theory tradition, as the experiences and perspectives of marginalized peoples are often able to highlight the ways oppressive systems work and identify biases and blind spots that exist in dominant ways of thinking.[9]

I asked questions regarding their experiences on campus, their perceptions of the meaning of race on campus, and their experiences with racialized interactions online. I explored their reactions to racist attitudes posted anonymously online by fellow students and sought to understand how those events impacted their perceptions of race, racism, and the university's racial climate. These conversations followed an interview guide, but what I learned with each case or participant influenced subsequent interviews, as questions were modified or added to reflect what had already been learned.[10] Questions also changed over time to keep up with current events that shaped the racial landscape. For example, I added questions when protests in Ferguson, Missouri, shined the spotlight on anti-Black police violence after teenager Michael Brown was shot and killed in 2014. As the "Black Lives Matter" slogan gained prominence with these protests, national and campus-based conversations about race seemed to change. While the phrase "Black Lives Matter" has become symbolic of and synonymous with the broader Movement for many people, the Movement for Black Lives (M4BL) and Black Lives Matter (BLM) are distinct organizations. I also added questions when Donald Trump was elected in 2016 and seemed to shake the foundations of the nice, subtle racist language that had previously dominated the political arena.

Throughout the research process, a number of students wanted to stay behind after their interviews to continue our conversations about race. Several students indicated that they enjoyed the opportunity to process their experiences with race both in-person and online. In these situations, I sometimes turned the audio recorder back on if students began sharing examples or stories that were relevant to the study aims. In other cases, I simply included notes on the conversation with my other notes on the students' appearance, mannerisms, and demeanor during the interview.

There were also several moments throughout the data-gathering process where students shared stories that were sufficiently traumatic that I decided to pause the interview and offer to connect these students with campus resources, including not only counseling services but also administrators in the campus diversity office with whom I had developed a rapport and thought would be willing to pursue incidents of bias. In none of these cases did students elect to stop the interview. In several of the cases, the students expressed a willingness to get in touch with my contacts in the diversity office, but to my knowledge no students followed through with those connections. This may be indicative of how difficult it can be to respond to racism or navigate institutional processes around racism.

Upon beginning the interview process, I almost immediately began to uncover unexpected and theoretically surprising findings that changed the direction of the study.[11] As I have described, I went into the study looking to push back against contemporary theories of racism that assume a subtle, covert presentation of racist ideologies. This assumption is based on the prevalence of societal norms that make overt racist language or actions taboo, which can be muted in some technological contexts. How does the internet affect the presentation of racial ideologies, and how do these distinct presentations affect the way students of Color think about race, racism, and interracial interactions on campus? This is the question I seek to answer in the first half of this book.

While I did hear about student experiences with more explicit presentations of racism in online spaces, this idea, that theories of racism cannot account for race on the internet, was far from my most interesting finding. I also found that despite being exposed to brutal language in online spaces, students of Color did not shy away from online environments. They didn't, like my cousins and I, *take off their headsets,* making the choice to avoid harassment by disengaging from the online space. Instead, they used online social media to construct spaces of critical resistance. The second half of this book explores these new online-based methods of resistance to racism and their influence on both online and offline racial discourse, racial consciousness, and online activism.

Throughout the analytic process, I engaged in a constant comparison between theory and data, taking note of the relationships between prevalent themes, existing social theories, and emergent theories.[12] In each chapter, I discuss and interact with research and theory that guide my interpretation and analysis of the data—whether gleaned from interviews, social media, or surveys—and make note of the places where surprising findings necessitate building new theories. The findings in this book deviate from contemporary theories of race and should shift the way we understand contemporary expressions of racism, how young people of Color respond to discrimination, and the innovative ways folks of Color harness technology to engage in acts of resistance. To say it another way, I am allowing the data to stretch the realm of possibilities, to shape the things we think we know but might not know as well as we think.

I published findings from the first stage of this project in 2017 with my doctoral dissertation, and later in a couple of academic articles.[13] But to make sure that what I was finding in Chicago was not unique, I expanded the study to other schools in other parts of the country, and to schools with different characteristics. Additional interviews and focus groups were conducted at public universities in Los Angeles and Atlanta, and private universities in New York and Boston, between 2017 and 2022. While the schools in New York, Boston, and Chicago are predominantly White, the schools in LA and Atlanta are more diverse, with the campus in Atlanta being over one-third Black and the campus in Los Angeles over one-third Latinx. Interviews from these additional sites add richness and complexity to the study and largely confirm my initial findings regarding the impact of online communication on campus-based racial discourse. To protect student identities, I don't mention the names of any of these universities, and instead refer to them by city throughout the book—Chicago, Los Angeles, Atlanta, Boston, and

New York. Similarly, all the people I interviewed have been given pseudonyms.

When I asked questions about online discussions about race, I let the people I talked to decide what places on the internet were most meaningful for them. Much of the research on race and social media seems to focus on Twitter (for good reason), as the platform has been central to largescale organizing efforts (and the data are easily accessible for researchers). But the students I talked to rarely discussed Twitter. They were more likely to describe experiences with race on Facebook, Instagram, Snapchat, or blogging platforms like Tumblr. By learning from my interviewees about the online spaces where they witnessed and engaged in race-related conversations, I was exposed to types of racial discussions that do not, like trending hashtags on Twitter, make the news but are still hugely important to the everyday lived experiences of people of Color.

In 2021, 69 percent of adults in the United States used Facebook, 40 percent used Instagram, 25 percent used Snapchat, 22 percent used Twitter, 21 percent used TikTok, and 18 percent used Reddit.[14] Adults aged 18–29 were more likely to use all these platforms, with 70 percent on Facebook, 71 percent on Instagram, 65 percent on Snapchat, 42 percent on Twitter, 48 percent on Tik Tok, and 36 percent on Reddit. The age differential is smallest for Facebook, which many young people describe as being for their parents. Black people also have a higher-than-average usage on all these platforms except Reddit, and Latinx folks have a higher-than-average use on all but Twitter and Reddit.

Over the past decade, online activism and discussion of Black Lives Matter has transformed public discourse around race. Activists, thinkers, and antiracists have used online communication, social media, and Twitter, in particular, to highlight anti-Black vigilante and police violence and insist that these are not isolated incidents or the work of a few bad apples. Instead, they are understood to be directly tied to a broader system of oppression—the rotten tree of White supremacy that stifles the pursuit of happiness, liberty, and even life for Black people and people of Color. Antiracist counternarratives are becoming mainstream because of activists of Color and their allies using social media to force the issue.

Black Twitter is central to this story. Black people are the most likely to use Twitter at 29 percent (compared with 22 percent among all US adults), but Black Twitter is more than numerical representation. André Brock defines Black Twitter as "an online gathering (not quite a community) of Twitter users who identify as Black *and* employ Twitter features to perform Black discourses, share Black cultural commonplaces, and build social affinities."[15] Tressie McMillan Cottom wrote an article titled "Black Twitter is not a place. It's a practice," which talks about Black Twitter as an archive of public memory.[16] Mark Lamont Hill discusses Black Twitter as a digital counterpublic where "members actively resist hegemonic power, contest majoritarian narratives, engage in critical dialogues, or negotiate oppositional identities."[17]

Black Twitter has been an engine behind the Movement for Black Lives and countless hashtags that use technology to push the ugliness of racism into the spotlight, changing and elevating the conversations people across the world are having about race and racism. Sarah Jackson, Moya Bailey, Brooke Foucault, Melissa Brown, Rashawn Ray, and others have discussed how intersectional these discussions are on Twitter, as they highlight the ways racism intersects with other systems of power, like patriarchy, heteronormativity, or ableism.[18] Intersectional critiques and counternarratives are becoming mainstream because activists of Color and their allies use social media to redefine the problem and force the issue.

This book is driven by the conversations I had with people about their experiences online and on campus, people who were willing to share their time, brilliance, pain, hope, resilience, and resistance. Still, the puzzle would be incomplete without also paying attention to how online conversations about race are changing the racial landscape.

This is why *When the Hood Comes Off* also looks beyond the college campus.

I use social media analytics to track trends in Twitter-based racial discourse between 2011 and 2021, demonstrating measurable changes in online discussions of race and racism over time. I combine a large-scale analysis of millions of tweets with a fine-tuned exploration of how people engage online, analyzing samples of social media posts that exhibit both the unmasking of racism and antiracist and resistance projects.

When researchers use data from multiple sources, we call this triangulation. Including social media analysis reveals that the things I found in interviews on the college campus also play out in other spaces. This book is the first study that combines in-depth interviews with both quantitative and qualitive social media analyses at this scale. I also include some findings from a nationally representative survey I conducted around how people experience online discussions of race, though only sparingly.

The book starts with a figurative pond, investigating how technology has changed the ways students talk about and experience race on college campuses. This is the focus of chapters 3, 4, 5, and 7. Then it moves from the pond to the ocean, applying and extending what I learned through interviews to the online space, with analysis of Twitter data in chapters 6 and 8.[19]

While these are vastly different types of data, they are pieces of the same puzzle, different arcs of the same story. Combining face-to-face interviews with social media analyses helps me truly bridge the gap between the digital and the physical and understand the hybrid, augmented reality of how we experience race and racism in the digital age.

This book represents nearly a decade of research, and in it I tell a story about how technology has changed the way we understand and respond to racism. This is a data-driven, human-centered story that asks questions of people and groups that are often ignored or undervalued. When faced with the ugly realities of racism, we may be tempted to take off the headset as a way of protecting ourselves from the horror. In this book, I'm asking you to keep your headset on, despite how hard it may be to stomach the discordant notes of oppression that twist our values, undergird our institutions, and hurt the lives of people of Color. Ultimately, I assure you, this book is hopeful, and points to the ways resistance can orchestrate change and expand our ability to imagine and demand a just society. My hope is that you, the reader, come away with an understanding of how online racial discussions have changed over time (and are still changing), and of the effects these changes have had on people of Color and our racialized world.

BOOK THEMES
The Hood

The Ku Klux Klan, with its white hoods and burning crosses, might be the most recognizable symbol of racism in the United States. The Klan is responsible for thousands of lynchings, beatings, shootings, and burnings of individuals, churches, businesses, and homes. The height of Klan activity took place during the Reconstruction and Jim Crow eras, a time when law enforcement was loathe to protect Black citizens from racist acts of terror.[20] Still, Klan members chose to hide their faces behind those emblematic pointed white hoods. The hood was protective: it concealed the identities of the farmers, store owners, officers of the law, and clergy members who wore it, which kept them from being held responsible for the destruction of life and property orchestrated by Klan activity.

For the past half-century, racism has been hiding behind a more metaphorical hood: subtlety. This book begins by exploring the decreased visibility of racism in the post–Jim Crow United States.[21] Today, formal discrimination is outlawed, overtly racist language is frowned upon, and racism largely operates through indirect and invisible mechanisms. This, I suggest, is racism "under the hood," or masked racism. This is the racism many of us have come to expect, the racism that scholars have spent decades uncovering, what David Williams calls the "everyday discrimination" that people of Color experience in mainstream environments.[22] It involves microaggressions that may feel too minor to be reported to HR, but have major impacts on the lives of people of Color, from health and mental health to experiences and performance at work and in school.

When the Hood Comes Off

Over the past decade, we have seen a resurgence of White nationalism—under the guise of the "Alt-Right"—which does not fit with the way scholars have understood racism to operate in post–Jim Crow America. Many of us have been thinking about how to combat masked, hidden, subversive, and systemic racism for so long that we forgot what it was like to have someone yell in our children's faces, "Go back to Africa!" Or, these days, "You will not replace us!"[23]

Perhaps that is why liberals could not figure out what to do with Trump except to call him crazy or a bigot. For many people who lean to the left, the idea that Americans would support Trump given his openly racist rhetoric was unthinkable. Nevertheless, Trump assumed the office of the president from 2016 to 2020, and many White supremacist groups united in their support of his administration.[24] During his term as president, Trump followed through on many of his far-right promises, and the fears that our nation would regress under his leadership were realized, at least in part. Trump may be out of the White House, but he's not done making trouble.

Fueled by everything from Trump to the Proud Boys, there has been no shortage of public events that seem to reflect more explicit expressions of racism becoming normalized. Anti-Asian attacks increased by 343 percent in New York City and 567 percent in San Francisco between 2020 and 2021.[25] Racist mass shootings have targeted Asians in Atlanta,

Latinx folks in El Paso, and Black folks in Buffalo, with the shooters in El Paso and Buffalo both developing their hateful ideologies online or using the internet to share their messages of hate.[26] In 2022, thirty-one members of a White supremacist group were arrested on their way to a Pride event with riot gear, while other counterprotesters had asked their allies to bring guns to the rally.[27] As White supremacist groups engage in racist, sexist, homophobic, antisemitic, and Islamophobic campaigns, Black feminist intersectional thought teaches us to analyze the interconnectedness of these domination projects that uphold the power of straight White men.

In a paper analyzing the ways racist groups have strategically used social media, Jessie Daniels quotes a White supremacist who says, "We've managed to push White nationalism into a very mainstream position" and that "people have adopted our rhetoric, sometimes without even knowing it."[28] Overt racists mask their openly racist ideologies to make their messages more palatable. As an example of this, Kathleen Belew suggests that during the January 6, 2021, Capitol riots White supremacist groups opted not to use racist chants or display open symbols of racism at the riots in order to make their cause more appealing to the general public.[29] White supremacist groups that have been on the fringe since the victories of the civil rights movement seem to be becoming more mainstream. How can we make sense of this apparent reversion to Jim Crow–era racism, from language coming from politicians to increases in openly racist acts of violence?

The current moment—with explicit expressions of racism becoming more prominent and more prevalent—can best be interpreted using a framework that centralizes technology as a mediator of ideas, human connections, and both public and private discourse. In this book I engage in a comprehensive investigation of the cause—how online communication changes racial discourse—and its consequences, or what these changes mean for people of Color. In chapter 2, "Once We Were Colorblind," I discuss the myriad tools we have to identify racism that is masked by subtly colorblind language or social structures, and I underscore the characteristics of online communication that can facilitate unique ways of talking about race, including the more explicit expressions of racism than we have come to expect in face-to-face interactions. This is racism unmasked. In chapter 3, "Mask On," I use interviews with college students of Color to highlight the informal rules of racial discourse that place limits on how folks talk about race in certain contexts, including norms of political correctness and a commitment to value-free academic discourse. In a sense, the rest of the book pushes back against this chapter, as technology renegotiates the rules of racial discourse. Chapter 4, "Mask Off," is a case study that explores the effects of an anonymous online student forum that amplified racist ideas and challenged the dominant modes of thinking about race and racism. Once racism is unmasked online, there are consequences in the "real," in-your-face world.

As I mentioned above, one unanticipated consequence of online racism is resistance. Without a hood to hide behind online, racism has become more visible than it has been since the civil rights movement, when images of nonviolent Black protesters being brutally attacked were seen on televisions around the world. As activists, scholars, educators, and citizens adjust their methods to fight against a more visible enemy, swaths of people are jumping on the antiracist bandwagon— people who might not have recognized the persistence of racism before it was exposed by viral videos, racist online posts, and increased White supremacist activity.

Several chapters focus on resistance. Chapter 5, "Digital Resistance," examines the huge differences between how folks of Color describe responding to racism in person, and how they describe responding online. In chapter 6, "Double-Sided Consciousness," I turn to Twitter and analyze tweets that use the unique spelling of White people, "Wypipo," and argue that this neologism is an example of Black folks on Twitter moving beyond W. E. B. Du Bois's "double consciousness" problem—which focuses on the inner mind of Black folks in the face of racism—and flipping the script by shifting the focus to problematic and oppressive manifestations of White supremacy. And in chapter 7, "Protest, Posters, and QR Codes," I discuss new forms of activism that are emerging due to the innovative practices of Black people and people of Color across the internet and on campus, and the effects these examples of activism can have on consciousness-raising and development.

The growth of online racism—and its apparent spread to the mainstream media and politics through Trump and the Alt-Right—throws a wrench in the ways many scholars—and the general public—have understood changes in racist ideologies in the post–Jim Crow era. *When the Hood Comes Off* extends these theories and reconciles our understanding of the subtle mechanisms of racism with more explicit expressions of racism online. In the final chapter, "Racism Is Trending," I analyze trends in tens of millions of tweets from 2011 to 2021 to provide a clear picture of changes in our collective racial, political, and social consciousness, and discuss the implications for resistance efforts.

From cell phone footage of police killing unarmed Black people to leaked racist emails and messages—or even comments from friends and family on our own social media—internet-based communication is exposing the continuing significance of racism in a world that has been pretending to be colorblind. While those who have experienced or studied racism have long known this to be true, for many the unmasking of racism is an eye-opening experience. How will they, and we, respond?

2

ONCE WE WERE COLORBLIND

But there are none so frightened, or so strange in their fear, as conquerors. They conjure phantoms endlessly, terrified that their victims will someday do back what was done to them—even if, in truth, their victims couldn't care less about such pettiness and have moved on. Conquerors live in dread of the day when they are shown to be, not superior, but simply lucky.

—N. K. JEMISIN, *The Stone Sky,* 2017

ADMITTING RACISM

There is a classic study conducted in the 1930s where a sociologist sent a survey to 250 restaurants and hotels across the United States, asking whether they would serve a Chinese person.[1] Over 90 percent of the restaurants that responded to the survey openly said they would not serve a Chinese guest. But six months prior, the sociologist had gone on an extended road trip to visit each of the 250 establishments with his friends, a Chinese couple. While the vast majority of these venues had stated that they would not serve Chinese folks as a matter of policy—they *admitted* they were racist—in practice, only 1 of 250 turned them away because of race. How ironic that restaurant and hotel owners were more willing to admit their racism than act on it.

Today, we seem to have the opposite problem. I remember learning this in my twenties, when I was visiting my cousins who lived in Bloomington, Illinois. It's a small town, surrounded by corn fields, and seemed like a different world for a Chicago kid such as me. While I'd often make fun of my cousins for living in the country, I really enjoyed my time there. Besides having a better gaming setup than I did (yes, same cousins), they didn't have a cramped city house like me—they had a sprawling house, newly constructed, with a trampoline in the backyard. I thought this meant they were rich.

We decided to hit the streets, and went bar and club hopping, visiting all five (yes five!) nightlife venues "downtown." The bouncer at the first club we visited, a White man, stopped us before we walked in. "No hats allowed," he said. CJ was wearing a Yankees hat.

"Damn, I forgot that," he said.

I noticed lots of people inside with caps on. "There are ten people wearing hats inside, right there," I said, pointing.

The bouncer looked back, and said, "Those are okay. They aren't flat brimmed."

"So the rule is no hats, if the brim is flat?" I was astounded. "Why, because this is how Black people wear hats?"

The bouncer smiled. "No hats inside, bro. You wanna come in, come back without the hat."

CJ took off his hat and aggressively bent it, the way we used to bend our hats as kids playing Little League baseball. I winced. That wasn't a Little League cap. It was a fashionable forty-dollar fitted hat—the kind Jay-Z made cool—and I didn't think getting into this club was worth destroying it, on top of the cover charge.

He put the hat on, looking ridiculous with its brim bent down like the top two sides of a triangle. "How about now?"

"Nope. It's still a flat brim."

We walked away, upset at the unfair treatment. CJ apologized, saying that he knew about the policy and forgot to leave his hat at home. I didn't think he needed to apologize. It was a bs policy and clearly biased. Across the street, we found a barbecue grill, which CJ opened and dropped his hat inside. "Don't let me forget it when we leave," he said. I remember hoping that CJ would forget his hat, which was now not only bent beyond repair but also resting on a stranger's greasy, crusty grill.

We went back to the bouncer, hat-free. He acted like he hadn't just sent us away. When we walked up, he stopped us from entering again. "Sorry guys, I can't let you in with those shoes."

All three of us were wearing gym shoes. But this was not a classy place—not like one of the clubs in downtown Chicago where you knew beforehand that you had to wear dress shoes to get in. When I looked inside, all I saw was gym shoes.

"You're joking," I said. "Not one person inside is wearing dress shoes."

"Oh, you don't have to wear dress shoes. You just can't wear those," he said, pointing down at David's Jordan brand shoes.

"So all gym shoes are okay, except for Jordans?"

"And Air Force 1s."

This was sometime in the 2000s, before hype sneaker culture was as mainstream as it is now. Wearing Jordan's, Air Force 1s, or flat-brimmed hats was associated with Black, urban, hip-hop style. In fact, I remember being upset when Nelly, a popular rapper, came out with the song "Air Force Ones," because I thought it would make our culture—rocking Air Force 1s with the strap loose, searching for unique colorways—as mainstream as radio rap. I stopped wearing AF1s after that song came out.

That night it was clear that elements of Black culture—a certain style of hat, a certain type of sneakers—were being used to informally screen club goers, keeping the clientele exclusively White. And our experience was not unique. Research investigating racist nightlife policies finds that baggy clothes, Timberland boots, and white T-shirts often find themselves on the "no entry" list, and Black club-goers can even be denied entry for wearing the same shoes as their White friends who are admitted.[2] In our case, the bouncer didn't *admit* he was racist, but he selectively implemented policies to kept Black folks out. If you surveyed bar owners in Bloomington, I can't imagine any would admit their reluctance to serve Black people. This would be an admission of racism and a willingness to break antidiscrimination laws. But survey results be damned, they found a way around it.

Think about how different this form of racism is compared with racism in the 1930s, when a restaurant owner felt comfortable—or ethically and legally in the clear—answering a survey saying they were racist. This example is indicative of how the rules of racism (and the definition of racism, which I'll talk more about in a few pages) have changed in the United States over the past century. In the twenty-first century, racists typically don't admit their bias. But for some reason, most people seem to expect them to, hesitating to call anyone racist who does not proclaim it loud and proud. Commonsense views see racism as an attitude or intentional act. By this definition, even if someone does something racist, if they don't admit that they were being racist on purpose, the act can be explained away as a misunderstanding. It is only when someone admits racism that many people become confident that they are, in fact, racist.

One example of this is former CNN anchor Chris Cuomo trying to catch racists with a hypothetical trap. During Trump's first campaign for the presidency in 2016, Trump said he could shoot someone in the middle of the street and his supporters would remain loyal to him.[3] Testing this idea in 2019, Cuomo asked several Trump supporters whether they would continue to support him if he was openly racist. When Cuomo presented this question to Kris Kronbach, a Republican running for Senate, Kronbach insisted that he did not believe racism had a place in America, but also said, "I'd have to know who was running against him."[4] In another interview with Kayleigh McEany, then the White House press secretary, McEany refused to answer the hypothetical question, insisting it was an attempt to misapply the racist label to Trump. Cuomo responded critically, saying, "Why isn't your answer, 'No, I don't support racism'?"[5]

Many of Trump's supporters remained constant despite his being accused of dozens of sexual assaults, insulting communities of Color, mocking disabled persons, and even making false claims of voter fraud to finance the repayment of his campaign debt. Cuomo's question implies that admitting racism, somehow, might do what none of the above could and force Trump's supporters to abandon ship. Why? Because most Americans, on both sides of the political aisle, condemn racism—or at least, the popular understanding of racism as being intentional and overt.

Trump was famously slow to denounce White supremacy. He refused to condemn support from White supremacists such as the Ku Klux Klan's David Duke during his campaign in 2016.[6] Then, after the White supremacist rally in Charlottesville, Virginia, where a young woman was killed after being rammed by a car purposely driven into a crowd of counterprotesters by a racist, Trump insisted that "there were bad people on both sides."[7] And during the presidential debates of 2020, when asked if he condemned White supremacy, he told White supremacist groups to "stand back and stand by," something they interpreted as a coded message of support.[8] It took numerous tries by moderators for Trump to finally say that he opposes White supremacy. Despite all of this, Trump insisted during the debates that he is the "least racist person in this room."[9] Trump has probably been called racist more than anyone in history. Still, he doesn't admit racism.

Cuomo's hypothetical reflects what many people think about racism: *you're only racist if you admit that you're racist.* Without an admission of guilt, it is possible for Trump's supporters to deny he is racist, despite the hurtful language he has used toward communities of Color and his long history of discrimination against Black people and people of Color, both in and out of office. But if Trump were to admit he was racist, so the thinking goes, his racism would become tangible, moving from an alleged attitude to fact. With an admission, Trump's supporters could no longer excuse his racist acts, nor explain them as being unrelated to race.

According to this logic, admissions of racism can be, and often are, worse than racist acts. You can do racist things that deny opportunities to people of Color or reproduce racial inequality—treat people of Color badly, only hire White people, decide to move when a Black family settles in on your block, or tell your children not to date members of other racial or ethnic groups—but as long as you don't admit you're racist (or get caught saying something openly racist in private or online—an inadvertent admission), you won't be interpreted as one.

There's a big problem with this idea—that someone is only racist if they admit it. Over the past seventy-five years in the United States, racists have, for the most part, stopped telling us when they are racist. Survey results show that racial attitudes are drastically different than they were during the Jim Crow era. Whereas 95 percent of White people believed that interracial dating was wrong in the 1950s, only 10 percent thought it was wrong by the early 2000s.[10] While some of these changes may be due to changing racial attitudes, we also know that surveys looking to measure racist attitudes have had to modify their questions (something I'll discuss later) because respondents know better than to circle the "racist" option. During Jim Crow, explicitly racist policies were legal. In a world where restaurants could hang signs saying, "No Dogs, Negroes, or Mexicans,"[11] owners were more comfortable being honest about their racism on surveys, and people in general were less likely to be embarrassed to say racist things. Today, discrimination is illegal and racism is viewed as an ugly relic of the past. It follows that openly racist, discriminatory language is taboo.

Given this shift, most people use the R-word (Racist) only sparingly. Racism is thought to be rare, and the label *racist* is reserved for those who are *racist with a capital R*—Racist and proud. Openly White supremacist folks who want to convince the world that White people are superior are Racist. Those who still believe, in the twenty-first century, that the racial order was in better shape during the time of slavery or Jim Crow: Racist. People who agree with Hitler, that killing millions of Jews would solve the country's problems, are Racist.

But this type of old-fashioned Racism is out of style. In fact, it's so far out of the norm that limiting ourselves to this outdated definition can obscure contemporary forms of racism. It frames racism as a problem that normal people do not have to deal with, enabling them to minimize its effects and externalize the blame. Identifying outliers, extremists, and "racists with a capital *R*" as the true racists precludes the need to examine the complicity of ordinary people in maintaining a system of White supremacy.

As an example of how blaming the outliers affects our thinking, let's look at the Capitol riots when Trump supporters stormed the Capitol building to disrupt the counting of electoral votes on January 6, 2021.[12] Trump and many Republicans were quick to distance themselves from this event, and Trump insisted these extremists were not his followers. They hoped the public would separate rioters from the then-sitting president or the Republican Party (something that will be more difficult after a White House aide testified before Congress that Trump not only was in support of the riots but also sought to join them in person).[13]

But at the same time as the riot, a hundred Republican congressmen voted against the election results. One hundred! Which was a greater threat to democracy: this vote or the riot? Footage from Trump's second impeachment trial showed how close the rioters got to politicians, and the danger associated with this event should not be understated. But, still, I think a case can be made that a hundred elected officials voting to end democracy was more dangerous than a group of fools whose Whiteness (read: not a threat)—not their cunning, planning, or preparation—allowed them to cause so much chaos.

While the vote against certifying our democratic election results might have been the biggest threat to democracy that day, the terrorists who stormed the Capitol were the bigger story. In the same way, it's the everyday, lowercase-*r* racism that drives racial oppression, even though it's Racism with a capital *R* that makes the news. Both forms of racism are destructive, their effects durable across time, and both negatively impact people of Color and our society as a whole. But most people believe that open racism and open aggression hold more weight. This can put us in danger of ignoring the insidious effects of covert racism, as decisions (or votes) made behind closed doors shape the lives of people of Color in deleterious ways.

A more responsible definition of racism includes both subtle and overt manifestations. I take my definition of race from scholar Lawrence Bobo, who himself cites William Julius Wilson: "Racism is 'an ideology of racial domination or exploitation that (1) incorporates beliefs in a particular race's cultural and or inherent biological inferiority and (2) uses such beliefs to justify and prescribe inferior treatment for that group.'"[14] Note that this definition includes not just an attitudinal component but also a structural component: the *function* of a racist attitude is to reinforce racial inequality.

Racism is language, ideology, or actions that support and legitimate a system of racial inequality. Hostility is not a requisite when the function of racism is to maintain White dominance.[15] People can justify racial inequality and legitimate discriminatory policies and systems without using hate speech or even being prejudiced. But if their ideas about race reinforce racial oppression, those ideas are racist. We need to be able to detect and resist racism even when racists don't admit their true feelings, when people have no hate in their hearts but support policies or practices that hurt people of Color, and when it is reflected in behaviors, attitudes, and policies that are seen as "normal"—when racism is under the hood.

MASKED RACISM

In the 1930s, racism was clear and out in the open, admitted and not stigmatized. You knew it when you saw it, whether it was a sign telling you not to drink from a certain water fountain or leading scientists claiming that Black people were physically inferior and doomed to die out. Today, racism is often more subtle. Instead of "No negroes allowed" signs, there are carefully crafted (and subjectively enforced) dress codes. Instead of denying Black people the right to vote, in the name of election security racists make sure it's harder for us to get to the polls.[16] Instead of fighting for school segregation, racists draw school district lines to follow "natural" segregation patterns—and change the lines if folks of Color decide to move across them.[17]

Racism has done such a good job of hiding that many are fooled into thinking it doesn't exist anymore, or at least isn't very common. For some, this is because they don't want to admit that racism exists, because it would challenge their understanding of their country, and the belief that their social positions come solely from hard work, not unearned privileges. Others may legitimately believe that racism is dead. For them, this is because they have an image of racism in their minds that comes from videos of dogs attacking peaceful protesters in the 60s, or slave owners smiling over the bodies of mutilated Black people. If you believe someone must admit their racism to be racist, then you will see racism as something that only comes from extremists and outliers, and you may be unable to recognize the ways racism is . . . just normal. Open White supremacists are disturbing and pose an increasing danger to our society. But for a long time now, open racism has not been driving the rules, policies, and laws that limit the life chances and opportunities of people of Color in this country.

When racism is hidden, subtle, and embedded in seemingly harmless interactions or fair practices, policies, and institutions, it can be difficult to recognize. I call this *masked racism*. Racist Whites used to be proud of their racism. But now that racists have stopped telling us when they are racist, how can we identify racism?

A good analogy for this process comes from facial recognition technology. It is really convenient to have a phone that unlocks when you look at it (of course, there are a host of privacy concerns associated with this technology, but that's an issue for another book). But during the worldwide COVID-19 pandemic, when everyone was wearing masks, this facial recognition software was much less useful. For a while, at the beginning of the pandemic, when I was in a grocery store and needed to check the shopping list I kept on my phone, I would hold it up to my face, forgetting that my phone didn't recognize me with my mask on.

In a similar way, some people can't recognize racism when it wears a mask—when it's subtle. It can be easy to spot (and laugh at) microaggressions when Michael Scott is being insensitive about race, gender, or sexual orientation on the TV show *The Office,* but difficult to recognize the microaggressions every day in our own offices, classrooms, and homes, and the humorless ways they negatively impact people.

Thankfully, the mask of subtlety isn't foolproof. You can still unlock your phone with a mask on; you just may have to put in a passcode. Or, depending on what type of phone you have, you can use your fingerprint or a smartwatch. In 2022, Apple even added a feature enabling iPhones to recognize faces with masks on.[18] In a similar vein, there are ways we can recognize even cleverly masked racism. For people who have suffered from racism, we often know what it looks like well enough to be able to recognize it, even when it's masked. It may take a second for us to be sure, or maybe we need to ask a question or two to confirm it. *Why are you stopping me, officer? Oh . . . yes, I see . . .* We know it, in the end.

Those who do not have a personal history or relationship with racism, or are not able to recognize masked racism every day, can rely on the hard work of those who have devoted their lives to identifying and fighting against racism. Activists, scholars, and educators in many fields and disciplines have developed smart and useful ways to track racism even as it's masked in everyday institutions, policies, and interactions that the world sees as simply normal. These frameworks and tools are like a passcode, allowing us to recognize racism, even beneath its mask, or hood.

Identifying Masked Racism
SURVEYS

One way to measure racism is through surveys. This was easier before racism was masked, and racists were willing to be honest about their racism on surveys. Today, researchers have identified something called social desirability bias, which refers to the tendency for survey-takers to want to give the answer on surveys that they think will make them look best.[19] Because almost no one wants to be seen as a Racist (some White supremacist groups even claim to not be racist), this can make the unmasking of racism through surveys difficult. For example, questions measuring "Old-Fashioned Racism" were designed to detect the types of racist attitudes that were common during Jim Crow:

- · Black people are generally not as smart as whites.
- · It is a bad idea for blacks and whites to marry one another.
- · If a black family with about the same income and education as I have moved next door, I would mind it a great deal.
- · It was wrong for the United States Supreme Court to outlaw segregation in its 1954 decision.

As you can imagine, few people late in the twentieth century were willing to admit these types of openly racist attitudes. So researchers came up with a different set of questions that were meant to measure anti-Black attitudes, without making as many "don't sound racist" alarm bells go off in survey-takers' heads.[20] The Modern Racism Scale (MRS) asked questions that were similar in theme, but accounted for the unwillingness of survey participants to admit openly racist ideas—like saying Black folk weren't as smart. Questions on the MRS included:

- Discrimination against blacks is no longer a problem in the United States.
- Blacks are getting too demanding in their push for equal rights.
- Blacks should not push themselves where they are not wanted.
- Over the past few years, the government and news media have shown more respect to blacks than they deserve.

There were several iterations of this scale in the 1970s and 80s, adjusting for how respondents interpreted the items. Several decades later, some of them seem to ask questions that today we might consider to be overtly racist. But at the time of their publishing, these questions reflected a more subtle racist ideology. Rather than asking whether participants think Black people are inferior and unwanted, they asked how participants feel about Blacks gaining more rights. To uncover masked racist attitudes, questions focused on whether participants want to deny the reality of racism, and or if they believe that efforts for equal rights are problematic.

Today, even the MRS is less effective, with the "correct" nonracist answer being pretty obvious to most survey-takers. With fewer respondents willing to admit racism, some researchers have turned to alternative methods, including measuring implicit, or unconscious, attitudes.

IMPLICIT BIAS

Human beings naturally place new information into preexisting categories. Without thinking, we make automatic judgements about people we meet. These snap judgements have evolved as a way of helping humans determine who is in the in-group, versus who is in the out-group.[21] Are they dangerous? Will they be helpful? Should I trust them? Unfortunately, these processes that were meant to keep us safe can be hijacked by stereotypes, which occur when we attribute certain characteristics to an entire group of people (usually negative), or prejudice, "an overall negative attitude towards a group."[22] On an unconscious level, many people maintain biases with regard to race or gender that impact the way they evaluate and react to people who do, or do not, look like them. This is called implicit prejudice.

So when researchers find that survey-takers are unwilling to say, "Yes, I'm racially biased," another way of identifying masked racism is by measuring respondents' implicit or unconscious racial bias through laboratory experiments. Some of these tests are simple. Participants are shown pictures of people of various races and asked to rate them on a variety of measures. How attractive are they? How smart do you think they are? Would you want to be friends with them? When test-takers are significantly more likely to say good things about White faces than faces of Color, they're given a number indicating their level of prejudice.

Of course, test-takers figured this one out too. Most psychology experiments are done on White college students, and nowadays many students who know they are being tested for prejudice make sure they say "good," "pretty," or "smart," when they see a Black face, just as they do for the White faces. So researchers took the test a step further. Not only do researchers measure how respondents rate faces of different races; they test how long it takes for respondents to give that rating. If it takes longer for someone to say that they think a Black person is smart than to say a White person is smart, when all else is the same (outfit, setting, attractiveness, etc.), researchers have determined this demonstrates unconscious bias.

One of the most alarming experimental studies of prejudice I've encountered is a video game designed for police officers.[23] The computer game showed White or Black individuals holding either guns or nonthreatening objects. Participants were tasked with deciding to fire on suspects based on whether they were armed, and researchers measured differences in response times based on the race of the subject in the pictures. Results showed bias in response times, as the police officers who took part in the experiment not only made the decision to shoot Black characters with guns more quickly than they shot White characters with guns; they also decided *not* to shoot unarmed Whites faster than they decided not to shoot unarmed Blacks.[24]

When I hear about cops getting in trouble for making racist posts on social media, or secretly being a part of a White supremacist organization (something the FBI warned was widespread and a growing danger), I immediately wonder how many Black people were hurt, killed, or put in jail because of this racist cop that was allowed to work as a public servant for years or even decades.[25] I think about what the so-called Blue Lives Matter crowd might say—that he's a bad apple, not representative of the whole. Yet they remain unwilling to find or punish the bad apples (not to mention examine the tree and look at the problem at a more systemic level). This makes me justifiably angry. But it's not just those hidden-White-supremacist-bad-apples that are doing the damage. Research such as this shows us just how dangerous even implicit bias can be. If it takes the average officer a few milliseconds longer to decide that a Black man is safe—that's all the time it takes to pull a trigger, whether or not that officer is Racist with a capital *R*.

What do these types of studies, conducted in laboratories, have to do with racism in the real world? Subconscious or implicit racism, measured in a controlled laboratory setting, cannot necessarily predict rates of discrimination in the real world, outside of lab conditions.[26] But these studies can teach us how racism operates. If a person of Color is mistreated, it could be that the person on the racist end of things is an old-fashioned racist, someone who would admit that, yes, they are discriminating. But more often, the person who is behaving in a discriminatory way may not recognize it. The White woman clutching her purse as a Black man walks by could have progressive politics and actively work against racism on interpersonal and structural levels. But her subconscious bias influenced her automatic response to race (in this case, securing her belongings) in ways that do harm. The subconscious nature of prejudice, or bias, doesn't absolve folks from racism, but it does help us understand what racism looks like. It doesn't need to be overt.

RACISM IN PRIVATE

Another explanation for masked racism suggests that racism hasn't only become implicit or accidental; it has become intentionally hidden. This theory—two-faced racism theory—suggests that racists know that their biased attitudes have gone out of style and are careful to not say things that would get them into trouble when they are in public, but will be open about their racial hatred or bias behind closed doors or in spaces where they feel comfortable.[27]

The study that led to this theory gave White college students journals, in which they were instructed to daily record their conversations and activities. Leslie Houts Picca and Joe Feagin found that the conversations college students recorded about race were dramatically different in public spaces, what researchers called the frontstage, versus in private spaces, called the backstage. For example, racist jokes were much more common in closed, trusted spaces, than they were in mainstream, public campus spaces. White students were more likely to admit biased, unpopular beliefs about race or racial differences when they were in backstage spaces that felt safe.

Two-faced racism theory suggests that public displays of egalitarian racial attitudes are inconsistent with privately held beliefs. According to this framework, we may suspect that the decrease in open racism we see in everyday interactions may be performative, and the measured changes in racial attitudes—as we saw with the scales used to measure racism—may be explained by participants wanting to answer correctly, not honestly. The rules of society say that it is not right to be racist. Many people have learned to play by the rules in public, even if they don't agree with them. This makes our job, in understanding and identifying masked racism, a little more difficult. And as we will see, different rules and norms for talking about race in different contexts— like public vs. private, or online vs. offline—may begin to explain why online racial discourse seems so different from what we experience in person.

STRUCTURAL ANALYSIS OF RACIST ATTITUDES

The explanation for masked racism that I find most convincing focuses less on racism as an individual attitude, and more on racism as a structure that limits the life chances of people of Color. From this perspective, we understand racial attitudes to be more than individual dispositions: they are manifestations of the racial hierarchy. Remember Bobo and Wilson's definition of racism; the emphasis here is on the relationship between racial attitudes and racial inequality. We might think about the structural paradigm as being the inverse of suggesting that racism must be admitted in order to be real. From a structural perspective, the motives behind racism, whether conscious or unconscious, intentional or unintentional, or admitted or not, do not matter as much as their function in justifying racial inequality. People aren't racist for no reason. They are racist because racist attitudes legitimate and reinforce their racial dominance, both materially and symbolically. The style or expression of racism will match the trending racist practices or policies in any given time period or context.

Bonilla-Silva writes, "Racism is the product of racial domination projects (e.g., colonialism, slavery, labor migration, etc.), and once this form of social organization emerged in human history, it became embedded in societies."[28] This is something we can trace. For example, during the time of slavery, Black people in America were treated brutally. They were bought and sold like furniture or livestock, tortured as punishment and in the name of scientific advancement, and killed with impunity (unless the murderer was made to pay back the slave owner for their loss of property).

How could Americans have justified such inhumane treatment of other human beings? Racism. Slavery didn't exist because of racism; racist ideas were created to justify the practice of enslaving and colonizing other humans. Slavery was a profitable enterprise that allowed the ruling capitalist class to secure free labor, in addition to controlling a market of human beings as valuable goods. Racist attitudes—the idea that Black people were subhuman, could not feel pain, or were dumb and naturally suited to hard labor—made racist Whites feel better about the atrocities they were committing in the name of profit. This is a reality that has been intentionally hidden. The more we believe that racial classifications and hierarchies are natural and immutable, the less we question the ideologies, processes, and practices that continue to reproduce racial oppression.

Similarly, during the Jim Crow era, laws in America kept Black and White people separate through segregation, including unequal resource distribution in schools, healthcare, and unequal access to government programs, from federally backed home loans to the Social Security Act of 1935.[29] The Chinese Exclusion Act of 1882 banned immigrants from China and was not repealed until 1943.[30] When the law of the land was racist—privileging Whites over Blacks, or White immigrants over Asian immigrants—openly racist attitudes were absolutely necessary to justify those laws. Racist laws preventing Chinese immigration or African Americans from fully participating in society would have made no sense without the support of racist ideas. Segregation was seen as natural, God-ordained, and best for both White and Black people. Excluding Chinese immigrants was a seen as a way to protect

American culture and economy. Americans were not embarrassed to admit their openly racist attitudes when their attitudes matched the law of the land. Black people were seen as inferior, and Chinese people were seen as a danger to society—in order to explain why they were treated as such.

Today, Jim Crow laws have died and overt discrimination is no longer legal. But, of course, discrimination still exists. Schools aren't legally segregated and funded based on the race of the students, but through de facto segregation (based on informal, not legal, means), schools are just as segregated now as they were in the 60s.[31] And because schools are funded by property values, schools in affluent White communities are significantly more well funded that those in poor communities of Color. Black people are not subject to a legally distinct set of laws, but they are stopped by police and arrested more often, and given longer sentences than Whites who commit the same crime.[32] Hospitals do not openly deny Black people access, but inequality in access to healthcare, unequal treatment by medical practitioners, and the fruit of racism and racial inequality in other sectors have led to significant health disparities, including Black people being more likely to have miscarriages, heart attacks, cancer, and early death.[33]

Remember, racist attitudes serve a purpose: justifying and perpetuating racism. So what kind of racist attitudes are needed to justify the types of racism we see today, in which laws, policies, and institutions are biased against people of Color, without admitting bias? You certainly don't need ugly, open racist attitudes. In fact, those might reveal hidden racism, which is by definition taboo, and invite more widespread antiracist opposition. Instead, masked racist attitudes and language support racist structures yet leave room for people to believe that racism is passé, not something we need to actively fight.

As an example, let's look back at survey responses to questions about school integration. Between 1972 and 1986, support for integrated schooling rose from 84 to 93 percent—something that would suggest that racial attitudes were becoming more egalitarian. But during the same time, support for government desegregation programs dropped from 35 to 26 percent.[34] Lawrence Bobo uses the term *laissez-faire racism* to explain this discrepancy and refer to the way people may support inequality in principle but not in practice. The racial order is maintained without the need for overt racism.

Another example comes from what Leo Chavez calls the Latino threat narrative.[35] In 1965, there was relatively little undocumented immigration from Mexico, as around a half-million temporary work visas were given to Mexican migrants. In 1965, the Bracero temporary worker program was terminated, and the number of visas given to Mexican migrants decreased to twenty thousand. Massey and Pren argue that this shift in policy, coupled with the continued demand for Mexican workers, led to increases in undocumented migrants from Mexico, as the "legal" route to working in the United States had been taken away.[36]

In response to rising rates of undocumented immigrants, newspapers and magazines began using alarmist terms to refer to rising rates of undocumented immigrants, first in marine terms—things like *tidal wave* or *flood*—and then later in martial terms—words like *invasion* or the *need to defend our borders*.[37] This type of coded racist language to refer to Latin American immigrants legitimates the harsh treatment of undocumented migrants—from Trump's policy on family separation, with children being kept in cages, to Black Haitian migrants being hunted down and whipped by patrol agents on horseback at the border—insisting that these practices are not driven by prejudice, but by the need to keep America safe.[38] The language may not be explicitly racist, but it results in the perpetuation of unequal and brutal treatment of people of Color.

THE COLORBLIND MYTH

From kindergarten to fourth grade, my oldest son, Malachi, attended a public school in Chicago's Hyde Park, a Southside neighborhood home to the University of Chicago, where I was in graduate school. The school was diverse, with about half of Malachi's class being Black, with sizable percentages of White, Latinx, and Asian students. There seemed to be a large number of families with university connections, either graduate students like myself, or university employees who couldn't afford the $35k annual price tag on at the University of Chicago Laboratory School a few blocks away (or perhaps those who preferred public education regardless of cost). I noticed that while the younger grades were racially diverse, the older grades seemed to be less diverse and majority Black. I wondered why this was. Was there an exodus of non-Black families once children reached middle-school age?

One day, this came up in conversation with the parent of a child in Malachi's classroom. He, a White man, seemed flustered at the notion that something racially nefarious was going on. "Have you ever been to a school board meeting? Have you ever met the principal? You should. She's very colorblind. I've been very impressed."

I smiled at this. He was using the term *colorblind* as if this was an admirable way of viewing the world—not seeing race, just individuals. Many people have been taught that being colorblind is the ultimate aim in a world plagued by racism. Can't we get past racism, and just see one another as human beings?

But that's not how we should think about the term *colorblind*. One of my favorite scholars, Eduardo Bonilla-Silva—someone I've only met a few times but consider to be my theoretical uncle (especially after he bought me a whiskey at a conference)—argues that colorblind racism is the most common form of racism in America in the twenty-first century.[39] What is colorblind racism? In Bonilla-Silva's words, "it explains contemporary racial inequality as the outcome of nonracial dynamics."[40] The colorblind racist ideology frames issues of racial inequality in individual terms, rather than racial terms. How is this racist? Because it ignores racial inequality and legitimates racist systems that perpetuate inequality. Remember the definition of racism—an attitude or action that supports or furthers racial inequality. Someone can be racist without feeling an ounce of hostility toward Blacks or people of Color, so long as their actions keep racial inequality or oppression alive. Whether conscious or unconscious, well-meaning or not, attitudes are racist when they prop up and support institutions, policies, or decisions that hurt the life chance of people of Color.

This can be confusing for some. How can someone be racist, simply for saying that they choose to ignore race in their interpersonal relationships—because they think race shouldn't matter, and they don't want to discriminate?

As an example, think about the disproportionate amount of Black folks in prison, with Blacks being five times more likely to be imprisoned than Whites.[41] If someone says, "The law is colorblind; those people are in jail because they did bad things," what does this imply about Black people? It suggests that they are criminal, morally inferior, or have some cultural or at-home deficiency that leads to their disproportionate incarceration.

The same goes for the achievement gap in education between White and Asian students, on one hand, and Black, Latinx, and Indigenous youth, on the other. If you believe these differences are due to differences in work ethic, ethnic culture, or parents who don't care, then your attitudes about race—that some groups just aren't committed or hardworking enough— legitimate economic and education systems that we know to be racist, privileging some groups while failing others.

This ties into the last way we can identify masked racism: data. We know that judges give Black and Latinx people longer sentences than White people convicted for similar crimes and with similar criminal backgrounds.[42] We know that race and gender impact quality of medical care, with doctors more likely to refer White men for specialized care than women or Black folks with the same symptoms.[43] We know that Black, Latinx, and Indigenous kids are more likely to be suspended from school despite no evidence of racial differences in behaviors.[44]

There is plenty of proof that racism exists, and proof is another way of recognizing racism even when it hides behind a mask. It's important to understand these ways, not only so that we know racism when we see it but also so that we can help others to see racism.

RACE, DIGITIZED

Digital, online communication has changed, and is changing, the way humans communicate, work, and learn. Its full impact—on dating, relationships, learning, and politics—is beyond the scope of this book. My focus is the particularly strong effect the internet has had on the ways we talk about and experience race. There are several noteworthy characteristics of online communication that shape the style of racial discourse. Understanding these characteristics is important as we explore how the internet has fundamentally changed the way we experience race and racism.

We can mean many things when we talk about online racial discourse. Online communication can be text-based, when people are typing on a forum without sharing pictures of themselves. Or it can be image-based—think about memes, racist jokes, or videos that are posted and spread online. There are also webpages and special-interest sites, built to advertise, showcase, or instruct in some way. And then you have video games.

When the internet was new, some people thought that online communication would reduce racism.[45] Users would not know what race their fellow users were, so how could they treat them unfairly? Humans tend to categorize by racial or ethnic group based on visual clues like skin color or hair texture, and stereotyping based on group differences can be automatic and unintentional. Without the burden of these visual differences, it was argued that racism might cease to exist.

Boy, were they wrong.

First, research on the exact type of online, text-based special interest community they were describing showed that Black and White users got along perfectly—until Black users happened to mention their race in text communications.[46] After months of online group interactions without incident, users who self-identified as Black found themselves excluded from group discussions and activities. This discrimination took place even though the community interactions never included any names or visual components. Another study found that when an online video game gave players the option of playing with Black avatars, users with Black avatars reported negative gameplay experiences based around user responses to their avatars (even when there was no voice-chat function, as there was when I played *Halo*).[47] Lisa Nakamura's research and writing has been central to challenging the idea that the internet is colorblind and to helping us understand online visual cultures and the unique and sometimes unanticipated ways race is performed and experienced in different online settings.[48]

When people talk about race in person, they're dealing with other humans, and tend to be concerned with what others think about them, or whether they can get in trouble for what they are saying. This of course impacts the things that are said in public, in private, or at all. But online, there's potential for anonymity. Online users can create a username no one recognizes, and say any kind of crazy thing they'd like, with no real-world consequences. On Twitter, some of the biggest trolls—people who post outlandish or incendiary things, seemingly to make others mad—have profiles with made-up names or profile pictures that give no clues as to who is behind the account. I've seen established Twitter users, in response to these mystery profiles, say, "I'm not going to debate an egg." "Eggs" (a grayed-out sphere) are the default pictures for new users who haven't yet customized their profile picture, and often it is assumed that accounts that have not customized their profile pictures are either bots,[49] or people just looking to offend or not engage in any meaningful discourse. In other cases, White supremacist websites have posted instructions telling their followers how to make fake Twitter profiles, posing as Black people, in order to sow discord among Black Twitter users.[50] While many fake Twitter profiles impersonating people of Color have been suspended in the past few years, it is difficult to discern just how widespread this tactic is.[51] And we may never know who is beyond some of the most inflammatory accounts.

But sometimes we find out, like when users who have multiple Twitter accounts—one with their name attached, and one with a fake name they switch to when they don't want to be identified—make mistakes that alerts users, or the platform, to the fake account. This happened in 2020 when a White man who worked as a college professor posed as an immigrant woman of Color on Twitter, using the account to argue against feminist and antiracist policies and ideas.[52]

Something similar happened in the aftermath of the 2021 Atlanta mass shooting at two massage parlors that killed eight people, six of whom were Asian. The police captain responding to the horrific event suggested the killings were not about race, saying the killer was having a "bad day." Later, this police captain was found to have previously posted anti-Asian pictures and memes on social media, leading many to suggest that the bias he demonstrated online was surely reflected in his colorblind analysis of the murders.

Beyond the anonymity of a fake name or picture, the internet can also give people the impression they are more anonymous than they really are. This is called social anonymity.[53] We know that both social and technical anonymity can make people feel like their online actions are separate from real-world social consequences and norms. What is seen as normal and acceptable in human communication is different in different contexts. This means people might feel more comfortable sharing or saying things online that would be considered taboo in settings such as schools or workplaces. These changes in norms have a particularly strong effect on how we talk about race, because we have so many unspoken rules in place to limit the ways we talk about race in face-to-face situations.

Online, those rules are different. Often times, they are the complete opposite. It seems there is always a story in the news about a college student making a racist post online, going viral, and then being asked to leave the university. Why do people continue to post racist things online, when the consequences can be so severe?

Research on the effects of anonymity in online communication, whether real or imagined, tells us to expect increases in antisocial behavior. Hostile language—from insults to foul language—is significantly more common in online discussions than in face-to-face discussions. And internet users are even more hostile when communicating using a mobile device, versus with a computer.[54] This increased hostility isn't always about race; it's a general characteristic of online discussions. The lack of eye contact between individuals, the inability to see a person's face, to really recognize that it is another human being in front of you, decreases the level of empathy internet users feel for one another.[55] And without the nonverbal elements of communication, like tone or body language, for instance, it can be easy to misinterpret other users' intentions, or more difficult to make a point without being offensive.

When you combine this increased hostility in online discussions with racism—whew, can things get ugly. There are literally tens of thousands of racial slurs thrown around social media each day.[56] Some come from bots, some come from racists, and some come from trolls who just want to make people angry. Many internet users learn to tune them out, seeing racist speech online as just part of the landscape, like graffiti[57] on a train. It may not look great, but it doesn't slow down the train.

It's scary that there are real people behind those ugly words. We don't know who they are, but we have been able to connect this type of language to real-world outcomes and situations. For example, Seth Stephens-Davidowitz conducted a study that looked at how many Google searchers there were for the n-word by county, and found that counties with more n-word searches were more likely to switch from having voted Democrat for Kerry in 2004 to having voted Republican when Obama ran in 2008.[58] This fascinates me. We have no idea who was behind these Google searches, but we know it correlated with voters who had previously leaned liberal but decided to vote against a Black liberal candidate. And in one of my own studies, we analyzed 50 billion tweets between 2012 and 2017 and found that the number of n-words used on Twitter increased following the days where Barack Obama gave a speech. We call this *digital rage:* internet users upset with a symbol of Black advancement, President Obama, venting their anger online. For internet scholars, the rage of the Trumpers, like what was seen at the Capitol on January 6, 2021, was no surprise. It's been showing itself online for a long time.

Another unique characteristic of online communication is the way it encourages and enables user-generated content. Anyone can create a YouTube channel and post a video for the world to see in just a few minutes. You can create an account on Reddit and immediately find forums or threads dedicated to your favorite books, movies, TV shows, video games, or politics, where you can join conversations with like-minded users. While there are many benefits to this type of participatory culture, it also enables White supremacists to build platforms spreading their messages of hate. Open racists, who feel they have to hide their racism at work, church, or school, are able to join online communities where their views are not only accepted but encouraged. And they are exposed to even more racist ideas, including calls to action to protect Whiteness from phantom threats, such as those described in N. K. Jemisin's *The Stone Sky* in the epigraph at the beginning of this chapter.

Jessie Daniels's work has exposed the ways White Supremacist organizations have been targeting online spaces since the beginning, creating websites that spread their hate—even websites that don't look like they're racist.[59] For a while in the 1990s, martinlutherkingjr.org[60] was a White supremacist site that told a racist, false version of MLK's history. Daniels's interviews with children who visited the site showed that many of them could not tell it was fake. We'll never know how many people were influenced by the false narratives on that webpage.

In the past few years, there's been a growing amount of attention paid to an invisible internet-based racism: the bias implicit in the algorithms, search engines, and code that determine the ads we see, the information we have access to, and the structure of the webpages and apps that we use.[61] Safiya Umoja Noble and Ruha Benjamin (among others) have written about the structures that shape the internet—code, tech companies, and artificial intelligence—that can be as racist as the structures that shape our offline lives—schools, jobs, hospitals, and prisons. Benjamin calls this the *New Jim code*, and reminds us that the fight against racism in the digital age is not just interpersonal, or with the ugly racism with a capital *R*—it's with big tech policies and practices that reproduce racial bias in global technologies.

This type of racism, from my perspective, falls in line with other forms of masked racism—racism embedded in institutions, structures, and code that we are meant to trust. It is the job of activists, educators, and people who mean to resist racism to expose and challenge these structures. In this book I deal less with what Benjamin calls racist robots, the behind-the-scenes, invisible biases that shape the ways webpages, advertising, and social media work, and more with the ways people of Color perceive, interpret, and respond to racism they experience online.[62]

Online posts can sometimes give us a glimpse into what backstage conversations about race look like—or what racism looks like beneath the hood. Remember that many White people are more comfortable making racist jokes or statements in private than they are in public. Many social media users may post their private thoughts without realizing (or caring, in the moment) that this statement can have a real impact on their lives beyond the social media platform. In this way, the internet can be a fusion space—neither backstage nor frontstage; both public and private. Dana Boyd's book on how young people use the internet articulates this dynamic.[63] Boyd writes that in-person conversations are assumed to be private, and are only made public through intentional actions such as recording a conversation. But assumptions of privacy don't hold online, where, depending on the platform, public statements may be the default, even when users post as if they were only talking to their friends.

A DOUBLE-EDGED SWORD? RESISTANCE ONLINE

As a student in the late 90s and early 2000s, I was always told by teachers that I wasn't allowed to use internet sources. I always thought this was ridiculous. There were times I would use internet sources or Wikipedia to understand something for a school assignment, and only after having a good grasp of the subject would I begin to look for *real* library sources that I could cite for my assignment. I'm sure I wasn't alone.

Once, when taking an advanced statistics class in graduate school, there was a concept that no one in the class could understand, no matter how the professor tried to explain it. Eventually, the professor googled the term and pulled up its Wikipedia page. We read the first line of the article out loud, and at the same time everyone in the class said, "ooooooh," and then busted out laughing together. Here we were, PhD students at a world-class institution with several advanced statistics courses under our belts, learning from a famous professor who was an expert in stats, but the first line of a webpage we weren't allowed to cite helped us understand something that we just couldn't get from the textbook or lecture. This is not knocking the professor—who was great—it just shows how powerful of a tool Wikipedia, and other user-generated online content, can be.

The same technological and social characteristics that enable the expression of racist ideologies can also be used to build tools that can be used to educate, empower, and build more just societies. And obviously, not just Wikipedia. Online social networks, from blogs to social media, give citizens more control of information and facilitate knowledge generation and political participation outside of elite interests.[64] For marginalized populations, online spaces can provide an important opportunity for the production and circulation of counternarratives and alternative resistance strategies. When I first started studying race online, and was primarily looking for ways to understand online racism, I became drawn to the work of André Brock, who highlights the opposing side of this dialectic—online resistance—including the ways Black bloggers used technology to challenge mainstream narratives around Hurricane Katrina.[65] Allissa Richardson's work on Black mobile journalism also investigates the ways "raw reportage" can challenge racist narratives and document anti-Black police violence.[66] This is illustrated in the documentary about the killing of Mike Brown and the Ferguson uprisings, *Whose Streets?* Video footage from on the ground in Ferguson provides a very different account of aggressive police actions during the protests than did mainstream media accounts.[67]

These examples point to the diminishing power of media gatekeepers to control the stories that capture the national (or international) imagination. Before social media, activists often relied on media gatekeepers to tell their stories, and this shaped some organizing strategies. For example, during the civil rights movement, there were varying philosophies behind the nonviolence protest tactic.[68] One school of thought believed that the goal of nonviolent protest ought to be disruption, making racism inconvenient for Whites; the other believed that the goal should be moral change, convincing people to not be racist. Both motivations and strategies for change proved effective, and understanding these competing (or complementary) motivations can help us interpret contemporary social movement goals and strategies, and how technology can change them.

An example of the disruption school of thought is the sit-ins. Large numbers of Black people and their allies would descend upon a place of business—say a diner—that did not allow Blacks. With fifty people standing inside of a restaurant, it becomes impossible to do business. The goal of this form of protest is to demand a change in policy: let Black people in. The owner of the business does not need to be converted, or have their views changed from racist to antiracist, for this protest to be effective. If their business experiences enough disruption—if the protest keeps them from making any money for several days—they may decide to concede and serve Black people, regardless of their racial attitudes. Protests focused on disruption are less worried about changing racist attitudes, and more worried about changing racist policies and incentive structures. You can keep your racist attitudes, as long as you let me in—or pay me an equal salary.

The second school of nonviolent thought came from a belief in the morality of the American public. Regarding nonviolent philosophy Charles Payne writes, "As Gandhi understood it, what underlies the nonviolent attitude is a confidence that your ugliest enemy can change," and also quotes the statement of purpose of the famed Student Nonviolent Coordinating Committee (SNCC): "Through nonviolence, courage displaces fear; love transforms hate. Acceptance dissipates prejudice. . . . Mutual regard cancels enmity. Justice for all overcomes injustice. The redemptive community supersedes systems of gross immorality."[69]

This idea that nonviolent methods can lead to deep personal and attitudinal changes underlies the strategic choice organizers made to attract media attention and put the ugly, vicious nature of racism on display for the mainstream public to see—and reject. Organizers planned ways to get the media to publicize their stories, their activism, and the violence that was being used to maintain White supremacy. Videos of marchers being beaten by police and private citizens, attacked by police dogs, and sprayed with fire houses broadcast the ugly methods used to maintain Jim Crow and racial segregation, and they had an impact on how many Americans thought about race and racism. Those brutal images were distasteful and inconsistent with how Americans viewed themselves as the land of the free. The belief was that when people witnessed the horrors of racism, their attitudes would change and this would lead to changes in policy.

But organizers could not always determine what stories the media chose to highlight. For example, the media focused on charismatic men in leadership to the detriment of Black women who led and were central to the freedom struggle.[70] And because journalists were less enamored with Black Power philosophies than they were with visuals of violence in response to nonviolent protest in the South, many intentionally portrayed Black Power leaders (e.g., Malcolm X, Angela Davis, or the Black Panther Party) and messages in a negative light and spread narratives that created a false distinction between Black Power and the broader civil rights movement.[71] Organizers needed the media, and the movement certainly benefitted from media coverage, but ultimately the (White-dominated) press decided what stories to tell.

Social media has altered this dynamic, somewhat. Because of social media, organizers need be less reliant on traditional media gatekeepers when trying to publicize their messages or efforts. When enough social media users talk or share information about a certain event, it trends, whether or not the media gatekeepers want it to. Videos go viral without executive approval. New gatekeepers have arisen, of course, such as the algorithms and tech giants that can limit the ways social media tools are used or the audiences activists can reach, but the fact remains that more media power is in the hands of the people now than during the civil rights movement. In the protests after the murder of Mike Brown in Ferguson, Missouri, my friend and colleague Jacob Groshek analyzed the most influential users on Twitter, finding that Black activists were more influential hubs for information-sharing on Twitter than mainstream journalists or mainstream news outlets.[72]

The most prominent hashtag in online activism, or anywhere, is #BlackLivesMatter. The phrase *Black Lives Matter* comes from a Facebook post by Alicia Garza in 2014, in which she was reflecting on the murder of Trayvon Martin, and the feeling that Black lives did not matter to the legal system or the American public.[73] No hashtag has been used more in the past decade. Through social media this message, that Black lives matter, has spread across the globe. Some misinterpret it. Others may use it as a way of deflecting accusations of racism. I remember after the 2020 uprisings after the murders of Ahmaud Arbery, George Floyd, and Breonna Taylor, it seemed like every business on Newberry Street (the preeminent tourist shopping location in Boston) had a *Black Lives Matter* sign up in their window—as they were boarding up their stores to prevent looting. I wondered whether they were truly committed to a world in which Black lives matter, or if they believed the sign would keep their businesses safe. Either way, whatever the motivation of those who use the phrase, its widespread usage is emblematic of the transforming power the Movement for Black Lives is having on public consciousness around issues of anti-Black violence and the prevalence of racism in the twenty-first century.

The Movement for Black Lives, often equated with the hashtag or phrase #BlackLivesMatter, has elevated our national consciousness. But online discourses around race and racism are not limited to this hashtag. Rashawn Ray and colleagues investigated how using justice-focused hashtags can create online communities, and Mark Lamont Hill has written on how Black folks on Twitter challenge racism and create sites of resistance.[74] Sarah Jackson, Moya Bailey, Brooke Foucault Welles, Safiya Umoja Noble, Brendesha Tynes, and Melissa Brown and colleagues have explored how Black and Indigenous people, as well as women of Color, insist that resistance against anti-Black racism comes from an intersectional lens, and they take into account the ways that racism intersects with other forms of oppression and marginalization— including gender, sexual orientation, class, or religion.[75] Apryl Williams's research highlights the ways Black Twitter exposes racism through videos and memes of White people who call the police on Black people for no reason.[76] Millions of tweets have put the spotlight on anti-Asian racism in the United States, and social media users are constantly challenging normative modes of viewing race and racism. Given these findings, Tressie Cottom makes the argument that technology should be central to our study of race and racism.[77] And while André Brock's work has explored Black resistance in multiple contexts, his recent book examines "the ways Black folk use the internet as a space to extol the joys and pains of everyday life."[78]

How do we make sense of this double-edged sword? Technology has the potential to amplify racist messages of hate, and explicitly and ugly racist content is common and often unmoderated and unchecked in online spaces. Yet internet technologies have empowered a new era of activism, and #BlackLivesMatter has transformed our national consciousness, bringing increased attention to the persistence and brutality of anti-Black racism. With Black Twitter leading the way, BIPOC[79] folks have been using social media to bring mainstream attention to critiques of racist structures and power dynamics.

Far from an equal push and pull, I suggest that the magnification of this conflict between racism and antiracism will tip the scales in favor of efforts to resist racism. Laissez-faire and colorblind racists are able to deny their own racism because they dissaprove of overtly racist ideas and policies that contradict widely held American values. For Americans who can stomach the subtle racism that is the underfunding of Black schools, but not, say, the murder of Black children by police, social media has the potential to change the way they think about issues of racial justice, perhaps even leading them to recognize the connection between subtle and overt forms of oppression. The increased visibility of racism will not lead us into a postracial society. It can, however, expose racist ideas to more critiques and challenges, which may, in turn, increase the number of Americans who are aware of and act against racism in the twenty-first century. In the chapters that follow, as I investigate the effects of technology on how racism is experienced, understood, and challenged by people of Color, I will bring clarity to the lasting effects technology is having on racism, resistance, and society.

3

MASK ON

The Rules of Racial Engagement

> What you must learn is that these rules are no different than rules of a
> computer system. Some of them can be bent. Others can be broken.
> —MORPHEUS, *The Matrix,* 1999

CRITICAL RACE THEORY (CRT) is a framework used in the
legal and social sciences to understand the persistence of racism despite the legal victories of the civil rights movement,
and to highlight the ways racism is embedded in American
laws, institutions, and power structures.[1] For the past few
years, CRT has been making headlines. Not because of some
coordinated attempt by educators, activists, or scholars to
bring this theory to the mainstream, but because it has been
under attack from conservatives. This attack seemed to
begin with former president Donald Trump, who in 2020
banned the use of CRT in government diversity trainings,
calling them "unAmerican propaganda training sessions."[2]
By August 2021, conservative legislators in more than half of
all US states had made efforts to ban CRT from schools or
limit teaching on racism or sexism.[3]

It's not clear that the people working to ban CRT even
know what it is. Roy Wood Jr., a correspondent on *The Daily*

Show with Trevor Noah, features in a short video bit where he provides comedic commentary while playing clips of conservative news pundits and politicians giving conflicting definitions of critical race theory, none of which describe the actual theory.[4] It's possible the anti-CRT crowd have legitimately misunderstood CRT. But it's more likely that those who dislike the theory are uncomfortable with antiracist efforts to highlight the durability of racism, how its effects persist over time, and the ways policies and practices increase racial inequality. Others, of course, use fear of CRT to drum up political support. Remember, those who wish to preserve the racial status quo want to keep racism secret, or under the hood. I can't think of a better way to hide racism than to outlaw looking for it. Or to make the claim that investigating racism is itself racist.

Policy makers are not the only people uncomfortable with discussions about race and racism. This chapter is about how people talk—or don't talk—about race in face-to-face settings. My interviews with students of Color about how they experience racial discourse on campus reveal disturbing patterns and informal rules that keep race talk to a minimum, and limit the ways folks of Color feel they can engage. The rules that dictate offline racial discourse may not always be as obvious as the decision to ban critical race theory. But like the CRT ban, they are designed to silence folks of Color and allies who are bold enough to bring attention to the machinations of racism in the twenty-first century. What type of bans or rules are there in everyday conversation? How do people of Color interpret the rules for talking about race? How do they experience the penalties for breaking those rules? What are the effects of these rules on marginalized groups and institutional racial climates?

This is a book about how technology changes our relationship to racism. To understand this change, we need to first understand face-to-face experiences with race and racism—sans technology. The conversations I feature in this chapter highlight the complex and sometimes contradictory rules of face-to-face discussion of race and racism. These rules are neither exhaustive nor universal. They do, however, highlight specific dimensions of racial discussions in face-to-face contexts that set the stage for the rest of the book as I explore how technology changes these dynamics in unexpected ways.

Rule 1: Don't Talk About Race

In 2010, Arizona banned racial and ethnic studies from public high schools in the state.[5] Arizona's rationale was that these classes were unAmerican, racist, and would bring division where there was none.[6] Tom Horne, then-superintendent of public instruction, even cited his involvement marching with Martin Luther King, Jr., suggesting the ethnic studies classes were contrary to Dr. King's dream. This is a twisted interpretation of both Dr. King's goals and racial/ethnic studies programs, which in this case were designed to engage Latinx students by teaching racially/ethnically relevant histories while challenging the biased, Eurocentric histories they had previously been taught.

When state officials visited the class, they witnessed a discussion of the contradiction between the values of freedom written about in the Declaration of Independence on one hand, and the founding fathers themselves owning slaves on the other. When reporting back to the governing body about this discussion, they described students being taught that the founding fathers were racist, something that was seen as evidence of unAmerican content in the program.

As slave owners, George Washington and Thomas Jefferson oversaw the race-based torture of human beings, and Ibram X. Kendi notes that they both wrote letters articulating Black people as biologically, culturally, and intellectually inferior.[7] Surely, justifying the torture, rape, murder, and buying and selling of humans, because you believe them to be inferior, meets the definition of racist—even Racist with a capital *R*.

It has been argued that we can't judge these historical figures based on modern values, but it was clear to many in the eighteenth century that slavery was morally wrong. George Washington himself communicated this in his letters, writing that he believed slavery was an affront to God and discussing his desire to free his slaves.[8] Despite this moral and religious conflict, Washington kept his slaves for financial reasons. Along the same lines, as scholar Clint Smith describes his visit to Monticello, Jefferson's plantation, he writes that while Jefferson understood the horrors of separating enslaved families by selling them, he also saw slave commerce as a profitable enterprise and sold over a hundred slaves in his life.[9] These founding fathers knew slavery was morally wrong, and chose money over ethics.

The logic applied by the Arizona legislature—that teaching historical facts about the founding fathers is unAmerican—is consistent with contemporary anti-CRT sentiment. This logic is also seen in how some institutions collect, or don't collect, data on race. I once attended a presentation by a professor studying the effects of police violence against civilians. I was stunned when the statistical research did not even include race as a variable. When I asked why, the answer was that police departments did not supply the data on race, and pushing too hard for that data could hurt the researcher's relationship with police departments. A compromised relationship with the police would of course have negative effects on the researcher's career.

It is not this researcher's fault that the police protect racial data, and I understand they were making the best of the data they had access to. But research on police violence that doesn't include race is colorblind in all the worst ways. It hides the reality that experiences with police violence differ dramatically based on race. In 2014, President Obama formed a taskforce that, among other things, investigated police departments and made it harder for them to hide the race problem.[10] But Trump halted those reform efforts in 2017, ensuring that the federal government would no longer look for racism or anti-Black violence in police departments across the country. It's hard to find what's being intentionally hidden, and even harder to find things you stop looking for. Hiding enables and grants immunity to racialized police violence against Black and Indigenous people, as well as folks of Color, without holding police accountable. Without evidence, how can we prove that there was racism in the first place? Trump's actions granted racism another mask, just as the country had been starting to recognize its face, thanks largely to the sustained Movement for Black Lives, from Twitter to the streets.

Silence is one of racism's strongest allies. In each of these examples, people in power attempt to silence discussions of race and mask racism's standard operating procedures. If we look the other way, if we refuse to talk about racism, it is able to reproduce itself like an invasive species. Talking about and organizing against racism is like hunting for the "murder hornets" that found their way into the northwest corner of the United States in 2020.[11] Finding and destroying the hive prevents more hives from popping up. Ignoring the new species allows it to hide among us, each hive birthing another until the land is overwhelmed.

When I asked students how people talk about race on campus, Amanda, a Black student in Chicago, said, "They don't. Unless you are with other people who experience the shitty end of racism." The prevailing thought among people I interviewed is that White people prefer to avoid talking about race in school settings, and tiptoe around the subject when conversations about race are unavoidable. Daniel, also a Black student in Chicago, explains his experience talking about race.

> Never around the friends who are . . . not Black or not Latino . . . a lot of White students just don't see it as a problem . . . They think that you can move forward and try to mend whatever racial tensions our nation has been through. They think, okay, that means we shouldn't talk about race, that race shouldn't matter.

Conversations about race and racism can shatter the illusion that racism is a problem for the history books, not today, or that people and societies have moved beyond racism. It is true, race shouldn't matter. But it did, and therefore it does. Racism persists, and the effects of exploitive policies, from slavery to Jim Crow to mass incarceration, shape the lived realities for many folks of Color. As Daniel says, "Centuries of these tensions, you know, race is part of our identity. It's not something we can wipe away." Talking about race, for White students, is bringing up ugly history, or sad stories that ruin the mood on a campus they believe to be beyond racism. For students of Color, the historical context doesn't seem so far away.

Just as Arizona policy-makers believed that high school students learning about the history of racism created racial division, in Daniel's experience White students seem to think talking about race has the same effect. The CRT ban may be a partisan issue, with conservatives pushing for the ban, but the informal ban on race conversations is bipartisan. Silencing or avoiding conversations about race is not commonly seen as a racist act, and is instead the norm—or the rule—even on campuses that students describe as being overwhelmingly liberal.

Sara, a Black student in Chicago, discusses another dimension of this rule, saying, "People always want to look for the factor that it isn't racial, to like, 'Oh this is an economic issue.'" When racial silence is broken, deflection may be the first line of defense. *Surely you're not implying something as ugly as racism is at play here . . . there must be another way to think about this.* People and institutions are skilled at explaining racial phenomena without using race, thereby lessening discomfort.

This example reminds me of an episode of *The Good Wife,* where the main character, defense attorney Alicia Florrick, is defending a young Black child who threw a book a classmate.[12] Florrick accepted a deal from the prosecutor, but Judge Baxter rejected the plea bargain and sentenced the child to nine months in a juvenile detention center. Florrick enlisted help from a statistician in the law firm to investigate Baxter's cases, and found evidence of racial bias in sentencing, with Black children receiving more punishment. But this racial bias had only been going on since July of the previous year, not Baxter's entire career. After some investigating, Florrick found that a Black man had broken into Baxter's house and assaulted his wife that past July, which Florrick believed led to Baxter's bias.

After further statistical analysis, it was found that Baxter had begun giving out harsher sentences to children of all races after the July incident. Eventually the team uncovered that Judge Baxter was a gambling addict, and the July break-in was perpetrated by a bookie who was trying to collect on a debt. Baxter began sentencing more children to more time in jail in order to pay back his debt and finance his gambling addiction.

Throughout the show, the characters were in disbelief that a judge could be racist, and only Florrick was committed to prove them wrong. Initially, I was encouraged that this mainstream TV show would explore racial bias in the criminal justice system. But of course, racism could not be the final answer—it's much too outlandish to suggest a judge is racist. It seemed the show was suggesting that a more believable occurrence would be a judge literally selling children to prisons for profit—just not in a racist way. God forbid.

The episode made me think of Jessica Simes's book on the geography of mass imprisonment, where she discusses how hard scholars and policy makers have worked to explain away racial disparities in the criminal justice system.[13] These attempts are ultimately unsuccessful, however, as the data clearly points to racial bias at every level, from the police to prosecutors, judges and correctional officers, as well as construction of the laws and policies they follow. A racist judge is not outlandish. As Michelle Alexander demonstrates in The New Jim Crow, racist sentencing is the standard.[14]

If people of Color insist on discussing race or racism, which breaks the first rule of racial discourse, their reputations may suffer as they become seen as people who think everything is about race. Research finds, in fact, that people of Color often remain silent in the face of racism because they are worried about being thought to be too sensitive around issues of race.[15] Folks of Color often grow accustomed to the ways White folks deflect attention away from racism or racial differences.

The thing is, I don't know many people of Color who jump to racism as the first explanation for everything that goes wrong in their lives. Research shows that around half of Black people believe that Black social problems are caused by a lack of motivation, not trying hard enough, or that "many of blacks' problems are brought on by blacks themselves."[16] Black folks who are more segregated, less educated, and have lower paying jobs are even more likely to hold these views than their more privileged counterparts. Other research shows that Black people are more likely than White or Latinx people to blame Black people for their unequal positions in society.[17]

Everyone, Black people and people of Color included, often look for alternative explanations to racism when experiencing or witnessing potential racial bias. When folks of Color decide to attribute something to racism, trust that we have already tested other theories in our minds and are landing on the explanation we would least like to admit. It's the explanation that is the most disempowering and difficult to demarcate.

Because there is an informal rule against talking about race, bringing up race can be treated as an offense. Jerry, a Black student in Chicago who talks about the way White folk respond to him when he breaks the rules and begins discussions about race, says, "Some people actually get offended, bringing up race at [the university] and I'm like, 'Yo, this is an intellectual discussion.' I've heard people talk about all types of messed up things . . . but race is off the table."

Being offended at the inclusion of race in a conversation indicates that some White people feel implicated in these discussions. This reflects the normal way of seeing Racism in individual terms ("only bad people are Racist"), and not structural and historical terms, which recognizes racist ideologies and practices to be normal, everyday occurrences. Responding to Jerry as if he insulted them for bringing up race is a quick way to shut the conversation down by reinforcing the rule of silence. To continue would be to explicitly court conflict, the opposite of what is usually expected in a social interaction between friends. Jerry shares an example of his attempts to discuss race being shut down.

> I remember my first year, talking about reparations and people laughing at me at my lunch table . . . I remember they looked at me and they were like, reparations (laugh)—one girl, she hit me. She was like, "(laughs) You're so funny." I was like, (laughs) I can't believe this just happened to me.

This example reinforces the reason silence around race is upheld as a rule. Talking about race can implicate White people, practices, or organizations, and by extension may ask things of them they do not want to give up, from unearned privileges to reparations. In 2022, Harvard University pledged $100 million in reparations in acknowledgement of how the university benefitted financially from slavery both directly and indirectly.[18] Princeton and Georgetown have also announced reparations programs. Many more universities have benefitted from slavery, but may be loathe to give up resources because of this history. Race is more than a theory or topic of discussion. It is lived reality. It hurts, and it is emotionally heavy to reflect on or talk about. And, for those who believe in justice, racism demands action. The first rule of racial discourse is silence, because talking about race and racism is like a rabbit hole that may lead where somewhere folks in power don't want us to go: toward demands for social change.

Given the informal rule that limits racial discourse on campus, many students discuss these conversations moving to private settings. Phillip, a Black man in Atlanta, says of racial discussions, "Here it's more forum-based . . . When people talk about race it's more of an organized thing." Outside of these special events, Phillip believes the extent to which people talk about race depends on their friend group, as in, "A lot of people still today [don't] want to talk about that type of thing."

Phillip and other students seem to be referring to what the literature calls counterspaces—clubs, study halls, organizations, or informal peer groups that offer social support, space for resistance and cultural expression (and are sometimes formed in response to microaggressions).[19] One reason counterspaces exist, in fact, is because of the first rule of racial discourse: issues of racial identity and justice are not perceived to be welcome in mainstream campus spaces. These groups are important, and they would be valued even if racial discourse was more welcomed on campus. But as they are primarily populated by students of Color, the racial discussions that take place there have little impact on the norms of racial discourse on campus, and White students can choose to remain oblivious to race-focused discussions and the continued relevance of race in campus life and the world.

Rule 2: If You Must Talk About Race, Stick to the Niceties

There are few values or freedoms that are seen as being more American than freedom of speech. For many Americans, freedom of speech means we can say whatever we want, whenever we want, including perpetuating racist narratives and beliefs. To some extent, they are right. It is legal for racists to make websites or pamphlets communicating their views, or to hold marches or events. But you don't often hear about anyone attempting to take this right from them. Even when White supremacists combine their freedom of speech with their right to assembly, and even when these assemblies turn violent, still they are rarely denied their "freedoms." It has been well documented that White supremacist events are regularly given more security than events associated with Black Lives Matter or the Movement for Black Lives. For example, in December 2020, in Washington, DC, a Proud Boys rally turned violent, including multiple stabbings, but rather than the combative police action regularly seen in response to the Movement for Black Lives throughout 2020, the police seemed to protect the group.[20] A few weeks later, another group of far-right extremists broke into the Capitol.[21] There were arrests, days, weeks, and months later, once the Feds had time to put evidence together—targeted arrests of people seen on camera engaging in specific acts.[22] At protests for Black lives, Black folks are routinely grabbed and arrested at random and on the spot.[23] The point is that racist and even violent freedom of speech is often protected where antiracist and nonviolent freedom of speech is punished.

The right to freedom of speech doesn't mean what many think it means. Court cases over the past hundred years have identified several things freedom of speech does not protect, including actions that could harm others, such as shouting fire in a public place, threats, some types of vulgar materials, or speech in schools that threatens the educational community, from obscenity to promotion of illegal behaviors.[24] Still, proponents for freedom of expression vigorously argue against policies or practices that they feel limit, or censor, the right to speak one's mind.

The counter to freedom of expression is often political correctness, a pejorative term for discourse perceived to be overly careful in order to avoid offending others.[25] The term *politically correct* (PC) comes out of conservative critiques of attempts at regulating speech on college campuses in the 1980s.[26] In the early 1990s, President George H. W. Bush went as far as to describe political correctness as a form of McCarthyism (which refers to the Communist witch hunt that accused Communists in the United States faced during the Red Scare in the 1950s) threatening colleges and universities.[27]

PC discourse, especially around race, is common in college and university environments. This represents another rule for talking about race. It can't be offensive. Lawrence Bobo, James Kluegel, and Ryan Smith refer to laissez-faire racism as a "kinder, gentler anti-Black ideology."[28] Remember that the laissez-faire and colorblind racism frameworks explain how racism in the twenty-first century is maintained through the niceties, without any need for overt racism to justify the racial order. Overt discrimination is illegal, which means that overtly racist statements or actions are seen as taboo.

But not everyone agrees that PC norms are desirable. For example, research finds that some conservative students feel that their true opinions about social and political issues are unwelcome on college campuses.[29] On some college campuses, PC norms are formalized with speech codes that tell students the type of language that is okay or not on campus.[30] Many workplaces have similar codes of conduct, and even when there are no official rules about what can and cannot be said, most employees know that racially offensive language—especially overt racism—can get you fired.

Students I interviewed certainly talked about PC language as being the norm at their schools. Micaela, an East Asian student in Chicago, connects this PC norm to intelligence, saying, "Everyone is so smart that I think they are always politically correct all the time. And that was something that I was not, I wasn't used to coming into college . . . people think that they have to be 100% politically correct and not offend anyone."

Zeus Leonardo and Michalinos Zembylas use the term "White intellectual alibi" to describe "Whites' attempt to project a non-racist alibi," and the way "Whites have built anti-racist understandings that construct the racist as always someone else, the problem residing elsewhere in other Whites."[31] The idea is that being smart is associated with avoiding potential controversies and using safe language is consistent with the way many people view racists as being ignorant, backward, or uneducated. Micaela is describing White intellectual alibis that separate the norms on campus from popular conceptions of racism.

In another example Chris, a Latinx student in Chicago, adds to our interpretation of PC language, saying, "In person, people aren't going

to say things that they might feel. But, you can clearly tell that there's
. . . I won't say, racial tension, but there's some things people won't say.
Or they want to be very politically correct about saying things."

Students of Color are adept at reading between the lines, or understanding the meaning behind silences, discomfort, or overly cautious
language when discussing race. Chris suggests they can often tell when
White PC language hides an idea or feeling that is simply against the
rules for White people to share. Jelani, a Black student in Chicago, says
something similar, noting that "you can feel someone kind of walking
on pins and needles, in a conversation."

Different students have different interpretations for the cautious
way they see their peers behave when race comes up. Destiny, a Latinx
woman in New York, laughs this off, talking about some of her interactions with White folks on campus.

> Like some person trying to speak Spanish and they turn to me and, like,
> "Does this offend you?" And I was like, "No. You want to learn the Spanish
> language, go ahead" . . . If you call me a spic, then we're talking some-
> thing else . . . Sometimes the sincerely hilarious shit that White people do
> just to make sure that they're not crossing a line . . . I feel like the people
> that are most bothered by White people are White people."

Desmond, a Black man in New York, says, "A lot of the times it feels
so overwhelmingly positive that it seems like it's the cool thing to do
. . . You're cool if you can be able to, you know, not be racist (laughs)."
Part of what's "cool" when talking about race is to go beyond the basic
PC language—the type that keeps you from saying things that might
be offensive—and move onto the next level of PC language: antiracist
buzzwords. Ibram X. Kendi's best-selling books have helped bring the
term *antiracist* to the mainstream, as he first used it as part of his framework for understanding competing racist and antiracist forces throughout the history of racism in the United States, and then again in another
book where he discusses strategies for contemporary antiracist

actions.[32] Using terms such as "privilege," "structural racism," and "recognizing my own bias," are likely to indicate that the speaker is "woke," or aware of and beyond racism. The problem with this language is that sometimes students think once they understand these concepts (at least at a basic level), the work is done.

White folks may hope to avoid ever being called racist by demonstrating their mastery of antiracist or critical terminologies and externalizing racist behavior and policies as being far from their own ideologies and behaviors. How can one be racist when they have proved their antiracism through language that expertly identifies and condemns the machinations of racism? Using antiracist language, in some contexts, can be a form of image management, and this language can be used to deflect attention, or as a form of protection against potential accusations of racism.

We might expect PC language to be protective for students of Color, as it should reduce instances of overtly racist language. But noticing signs of insincerity when talking with friends can still be hurtful for folks of Color who believe that their peers are just following the rules, using the niceties and White intellectual alibis, instead of engaging with them honestly or deeply. Valerie, a Black student in New York, has an incisive perspective about the trendy nature of antiracist lingo.

> When I talk to white friends sometimes, they, like, they understand concepts of the structure of racism and stuff, so they'll drop like big words like, "Oh I know about, like, the prison complex," and stuff like that . . . And I don't think it really goes farther than that because they never lived through it . . . Conversation about race amongst the students is kind of, it's on a superficial level. Like, people know those terms but don't really know how to go deep.

I've been teaching classes on race for nearly a decade. One change I've noticed over time is that each year students seem better versed in their ability to articulate racism. When I first started teaching race, many students were astounded by our readings and discussions of masked racism and had no idea how insidious racism was. Of course, I still have some students who feel like a veil has been lifted from their eyes after taking the course, but these days it's more common to get students who come in with a pretty good knowledge of how racism works.[33] They understand structural and masked racism, and aren't prone to thinking that racism is dead, or dying.

One of the biggest challenges I face in teaching classes on race and racism, therefore, is not convincing students that racism exists, but figuring out how to get students to recognize their own biases. The Social Work students I teach are predominantly left-leaning and often come in with antiracist identities. They have invested their time, energy, and talents into dismantling racism before ever stepping into class. Sometimes they can even seem to have a better handle on intersectional issues of oppression than the faculty, and I am consistently impressed by and learn from their ways of thinking. On one hand, this is immensely refreshing: to teach students who don't need to be convinced that racism is a problem (this is *not* the case everywhere, in all programs). But it brings a new challenge. How do you convince socially aware students that they have more to learn about racism? How do you get students to interrogate their own biases, when they already speak the language of antiracism? How can we identify biases that are couched in the niceties, or in White intellectual alibis?

Students of Color I spoke with believe that their White peers operate according to the nice, PC rules of racial discourse, not saying anything offensive, and saying the right, smart thing, in order to maintain innocence in racial matters and avoid deeper conversations. When students of Color recognize/suspect these discursive strategies, they can make judgments about the motivations of their White peers. Without talking to White students, it is impossible to state for certain what attitudes or ideologies drive this discursive tiptoeing. For some White people, these strategic moves could be hiding racism. In other cases,

White people who identify as antiracist may use this language in honest attempts to articulate the many machinations of racism and engage in antiracist practices in their lives. It is possible, even probable, that some students of Color misjudge some White peers. Whether or not their judgements are accurate, however, they tell us something of how students experience racial power dynamics on campus, and how they interpret antiracist performances from their White peers.

Rule 3: White People Are Allowed to Break the Rules

When I was a graduate student at the University of Chicago, I spent five years working as a resident head[34] in the residence halls. My first year on the job, the director of the Housing Department was a Black woman. In my first training, she told me about her "no nonsense" policy regarding the n-word. Students who used the word were kicked out of housing, with no exception. To emphasize how serious she was about this rule, she told us about a Black student who had been kicked out of the dorms for using the n-word publicly and in a joking manner. This surprised me. Many Black people use the n-word colloquially, and this is typically accepted as being different than the racist use of the word, as the n-word carries a different meaning when used by Black folks than when it is used by White folks—than again when it is used by non-Black folks of Color. Ta-Nehisi Coates gave a good example of how this works, suggesting that while it was appropriate for his wife to call him honey, strangers on the street calling him honey would be uncomfortable.[35] Still, I understood where my boss was coming from, because the university might not recognize how the same word carries different meanings, and there could be legal problems behind unequal or "biased" reinforcement of the rules.

Each year, when I talked to the group of between sixty to a hundred incoming freshmen students on my floors, I communicated this zero-tolerance policy. But a few years later, when we had a new director of

Housing—a White woman—our annual training communicated something different about freedom of speech. The new director told us we should hold meetings with students who used hateful language to help them to understand why it was problematic, not kick them out.

I raised my hand to ask a question, "I thought we had zero tolerance for racism in the dorms, and that students were asked to leave when they made these types of statements?"

"That has never been a policy, to my knowledge," the new director of Housing said. She looked around at the people presenting from the legal department, who shook their heads.

I followed up with administrators and other resident heads to make sure I hadn't imagined this policy. A few colleagues also remembered the policy to kick students out for using hate speech, but didn't know when it had changed. This made me suspect that the Black director had made her own policy when it came to racism. With nothing on paper, administrators can be forced to make decisions on a case-by-case basis. People concerned with reducing or eliminating racism may enforce rules against racist speech, but those who see racism as being a minor, or peripheral problem, might be less inclined to make or enforce those rules. Those inclined to use their power or authority to eliminate racism in the dorms can make them safer and healthier space for students.

This made my old boss a hero in my eyes. It reminded me of what Michael Lipsky calls street-level bureaucracy, which refers to the ways public service workers, from teachers to social workers, are able to use discretion in how they interpret and enact policies from higher up, even if their jobs don't explicitly give them this authority.[36] I decided to follow her example and act like a dorm-level bureaucrat. I had limited power, but did have some discretion over the rules I emphasized or enforced. When I gave the introductory talk to my students that year, I still told them that there was a zero-tolerance policy with regard to racist speech. Even though the higher-ups might not have had my back, that's not something I was going to play around with. I was raising Black children in an apartment in those dorms, and while I expected my kids to hear more curse words than the average child because of how much time they spent with college students (they also overheard more conversations about science, philosophy, and pre-med life, and I was okay with this trade-off), I'd be damned if they were going to hear the n-word tossed around by White students without consequence.

Some might assume that using the n-word—which has been called the "the nuclear bomb of racial epithets"– would break the second rule of racial discourse, which says that people, including White people, must adhere to the niceties when discussing race.[37] But rule 3 says that White people may break the rules. In White-dominated spaces, there is often room for White folks to operate as race-bureaucrats, using their discretion as to when typically inappropriate ways of discussing race can be deemed appropriate on a case-by-case basis.

For example, Rachel, a Black woman in Chicago, told me about a White student who made racially insensitive comments about Black women when playing a word-based board game. When Rachel called him out for these comments, things went a step further.

> And he says, "well, just wait 'til another word comes up, because I'm just gonna say, like, nigger, nigger, nigger." And he just said the 'N word' like five times, and I was just . . . I was really upset, so I just got up and left and went to my room. But, I think the worst part of that incident was, because I mean people are going to be stupid regardless, but I went to report it . . . to my [hall director], and they basically just encouraged me to like talk it out with him. And I was just, like, I don't know.

In this example, the student felt free to break the rules of racial engagement because, in his view, it was necessary (or at least his prerogative) in the context of the word-association board game that was being played. As a result of this confrontation, and the lack of response it received, Rachel moved out of her dorm. This was not an isolated incident. Rachel told me about another Black woman who was called the n-word directly to her face, and also told her respective hall director about the incident. This hall director told other students in the dorm about the situation, and the Black woman who was verbally attacked was ostracized by other students for "making a big deal" out of the incident and ended up leaving the dorms as a result.

I couldn't believe what I was hearing, and stopped the interview for a minute in order to offer to connect Rachel with a contact in the Chicago diversity office, someone I was confident had the motivation and capacity to call to task both the offending students and the unresponsive hall directors. Rachel seemed open to the idea of meeting with this contact, but the meeting never happened, and in this case rule 3 was upheld. For students of Color, not only can finding administrative support seem improbable, but approaching unknown allies can also be difficult on an interpersonal level. Students of Color are all too aware of the potential educational and social costs of pursuing justice. People in power may be more likely to uphold rule 3 than to hold White folks accountable. Of course, if this incident were to have gone viral, I'm sure it would have turned out differently—but this chapter is about race on campus *without* technological intervention.

Amanda gives an example of another racial experience that was more explicitly problematic, but still not something she felt she could respond to.

> I was in the dining hall one time and these White kids were talking about Chief Keef[38] and this White boy said "nigga," under his breath, but looked directly at me . . . I'm one person in a table of White boys. If I was, like, "Hey that's not cool don't say that," all of a sudden then I'm labeled as this crazy angry Black woman . . . You have to pick and choose your battles . . . The fact that he said it was just alarming to me, but he looked me in the face . . . I knew that he knew what he was doing was wrong especially in my presence . . . There are just like tons of other microaggressions that I just encounter everyday.

Whether use of the n-word can be classified as a microaggression is up for debate; most of the time it's thought of as a racial slur, or hate speech. In this case, the student who said the n-word was reciting a song lyric by Chief Keef, a popular rap artist. Amanda mentions that the student made eye contact with her, which she believed indicated that he knew he was saying something taboo. In fact, this reminds me of my example in chapter 1, when my friends glanced over at me when they decided not to say the n-word along with the song. Still, the student may have also thought he was falling short of using hate speech because he was quoting the lyrics of a song, similar to the way a student may feel comfortable saying the n-word while reading Mark Twain's *The Adventures of Huck Finn* out loud in class.[39] Both reading aloud in classrooms and quoting rap lyrics might be viewed as acceptable situations that allow for the breaking of the rules of racial engagement and use of a racial slur on campus.

Amanda knew this student would get away with breaking the racial rules of engagement and believed that she might face negative consequences and be labeled a "crazy angry Black woman" if she confronted him. In an essay about women of Color and fury in response to racism, Audre Lorde argues, "Any discussion among women about racism must include the recognition and the use of anger," and, "Anger is an appropriate reaction to racist attitudes."[40] That Amanda felt she must avoid getting (justifiably) angry at arguably the most insulting word in the English language demonstrates the injustice and uneven application of social rules around discussing race. Rule 4, in fact, is that Black people must not make White folk uncomfortable, something we'll explore more soon.

In September 2016, the Dean of Students at the University of Chicago sent all incoming students a letter explaining the university's policies regarding freedom of expression.[41] He explained that the university's commitment to academic debate meant that it would not support the creation of intellectual "safe spaces," or spaces free of content that could be offensive to students. Then, during orientation week, all first-year students were required to attend an Aims of Education address where the faculty speaker explained that while students could disagree with peers or speakers, they could not expect the university to punish those people.[42]

The publication of this letter sparked a national debate around the perceived conflict between free speech and campus speech codes. Over the next five years, colleges and universities across the country contributed to the conversation.[43] What is often implicitly communicated to students of Color is that colleges and universities will be loath to intervene in situations where students encounter racialized harm. While higher education institutions frequently respond to overt hate speech with statements about their commitment to diversity, there is less precedence for them to respond as openly to racial climates rife with microaggressions and layers of stereotypes.[44]

When schools across the country send messages that intellectual freedom is more important than eliminating harm, we should not be surprised when they bleed into classroom and campus dynamics. Rebecca, an Afro Latinx student in Chicago, says, "In an academic environment . . . everything suddenly seems legit and it's just, like, 'Yo, we're talking about this an academic way, anything flies, this isn't about sensitivity. This is about us discussing race.'"

There is a sense that White students do not tread as carefully when talking about race in the classroom as they might in nonclassroom contexts on campus. Academic inquiry or classroom discussions delineate more exceptions to the rules of racial engagement. Ideas that may be offensive in other contexts can be expressed as (supposedly) dispassionate discourse. But even when (or especially when) White students talk about race in a purportedly dispassionate manner, they betray a callousness to the fact that their comments deal with the lived realities of students of Color. A personal component is unavoidable for students of Color. Without proper facilitation, these types of discussions leave students of Color feeling as if their classroom was a hostile territory.

This book explores the way context changes the types of racial discourse that are considered acceptable. There are differences in accepted norms inside and outside the classroom, in person and online. When White students make assumptions or generalizations about race that go unquestioned in a classroom, students of Color are stuck in a perpetual bind; speak up and risk offending their peers, or remain silent and suffer abuse. This brings us to the fourth rule.

Rule 4: Don't Make White People Uncomfortable

In my first class as a PhD student, my mentor, Professor Charles Payne, asked the class a question, "What are the lasting effects of slavery?" For the next ten to fifteen minutes, everyone in the class—all first-year PhD students, eager to prove our knowledge—demonstrated how much we knew about the history of racism and the oppression of Black people in America. Educational inequality; discrimination in the unions; unequal access to resources; soul wounds. I remember looking up at the board and feeling like we pretty much hit it all. And I was proud of my own contributions. That was the point, right? To show that we had read the right books and understood that racial inequality was not the fault of Black people. It was by design, a legacy of historical and contemporary oppression.

Dr. Payne wasn't so impressed.

"It's interesting," he said, "that every single thing you have up here is about the effect of slavery on Black people. I didn't ask you the effects of slavery on Black people; I asked you the effects of slavery. To suggest that slavery only impacted Blacks, and not Whites—that Blacks are victims with no agency—isn't that a backhanded way of saying that Whites are superior? They are acted on but don't act?"

We were silent. It seemed we had missed the point of the exercise. I confess, I still didn't get it. Then Dr. Payne gave us a second question, clarifying what he meant: "What are the lasting effects of slavery on White people?"

The point here was that slavery has had a lasting effect on White Americans, just as it has had a lasting effect on Black Americans. For some White people, the legacy of slavery makes it difficult to understand the connection between historical and contemporary forms of racism. White folks often see slavery as a stain on American's past, a blip they wish to both minimize and distance themselves from. If I had one bitcoin for every time I heard, "That was centuries ago; it has nothing to do with me," or, "I've never owned slaves; don't blame me," I could buy a Tesla for each of my friends. Being defensive can take away one's ability to think about race in historical and analytical terms, instead of individual terms. One of the legacies of slavery is that it hurts many White people's ability to talk about race, as they attempt to deflect blame away from themselves; their wealth, privileges, and accomplishments; or even Western society.

This tendency for White people to get defensive when discussing racism is key to another of the rules about racial discourse: don't offend White people. If you do blame White people for racism, or racial inequality, you might be accused of reverse racism. Because most people define racism as the ugly, overt style of racism that was common before and during Jim Crow, the implication that current society, or regular individuals, could be racist, is offensive for many. Robin DiAngelo, a White woman who travels the country giving workshops on antiracism, talks about what she calls White fragility, or the tendency for White people to push back against the discomfort associated with talking about race, thereby focusing discussions of racism on themselves and how they feel.[45] Many in-person discussions of racism are shaped by White fragility and are designed to make Whites feel comfortable. When I talk about racism with White people who don't think about racism often, I try to get creative to make sure that they don't feel personally implicated, so that we can actually have a discussion. For those who think they *should* feel personally implicated—that's an Advanced Racism course; I'm talking about Racism 101.

Trouble can arise when people of Color break the rules and cause White discomfort. For example, April, a Black student in New York, says, "Sometimes when I'm frustrated or when I'm angry and I'm tired of living this experience and having to operate in this space, I don't want to . . . consider your feelings when I'm the one being oppressed, you know? . . . I'm just like okay, say what you want to say. I'm going to roll my eyes." April's decision not to respond to racially problematic discussions not only preserves her personal and emotional energy, but also protects her from the social consequences of making White folks uncomfortable. In April's words, "I cannot afford to have someone hate me for two years straight."

Folks of Color are careful not to offend White folks for a reason. There is a long history of White people taking retribution against people of Color who upset their sense of racial equilibrium or simply make them uncomfortable. From firing people from their jobs or punishing them in school, to socially ostracizing them or even physical violence—there are real-world consequences to breaking the rules and making White people uncomfortable.

One of the most obvious examples is the way Black folks are taught to be careful around and respectful to police officers (i.e., "the talk"). If an officer senses disrespect from a Black person, they tend to get vindictive and potentially violent. Sometimes it feels like they are just looking for a reason. As an undergrad in an all-White town, I was pulled over by the police so regularly that I had a routine: I would hand the officer both my license and my student ID, and explain that I was driving to or from campus to study. I would always see the officer loosen up once they realized I was a college student from down the street. All of the sudden I was perceived as safe. The one time I forgot to do this, the officer who had pulled me over grabbed her gun. I'm ashamed of this routine and shouldn't have had to act this way. I'm always flabbergasted when I hear how comfortable some White folks are talking back to the police.

The stakes are even higher for women of Color. As Audre Lorde discusses the limitations of anger, she says, "For women raised to fear, too often anger threatens annihilation. In the male construct of brute force, we were taught that our lives depended upon the good will of patriarchal power."[46] Women of Color must worry about the potential ramifications of their actions upsetting the status quo of multiple hierarchical systems at once (i.e., White supremacy and patriarchy, among others). For example, Tameka, a Black student in Chicago, talks about a class discussion and responding to another student that compared slavery in ancient Greek civilizations to slavery in the United States.

> He said something that I thought was just way wrong and I—I raised my voice. I didn't yell, but I raised my voice to match his tone. I looked him in the eye, I was, like, "This is not correct. You are profoundly misunderstanding this book," and, he's like, "Wait, I don't want you to get angry, I don't want you to turn into an angry woman. Just calm down, just calm down." And I was kind of like okay, so I'm having an intellectual argument with you, but then you disqualify my thoughts and opinions because you want to automatically subject me to the typical angry Black woman that you see on TV shows? . . . I was just kind of shocked—I couldn't even finish my thought.

This is the exact response that other students, including Amanda a few pages back, seek to avoid. Tameka's comment broke the rules, was perceived as a challenge to both Whiteness and masculinity, and was deflected by an accusation that was overtly gendered, which Tameka perceived to also have an unspoken racial dynamic given the prevalence of the angry Black woman stereotype. Tameka continues to share how she responded in this case.

I was kind of silent for a while because . . . my typical reaction in that kind of situation is to kind of go off. I purposely did not want to become the angry Black woman that he had just accused me of being, so I sat quiet so I could find my words and we kind of moved on to the [next] subject. So I raised my hand and I was talking about the subject, but I kind of made a snarky comment—I was, like, "Oh, maybe I should whisper, because I don't want people to think that I'm angry right now."

In modifying her behavior in response to the man's weaponized use of gendered and race-based stereotypes (something Lorde would likely see as being worthy of righteous anger), she reified his position, as if her silence was necessary to bring the classroom back to dispassionate equilibrium. It is significant that this interaction took place in a classroom taught by a woman of Color. While the professor did not intercede, she laughed when Tameka offered to whisper in order to be accepted as an intellectual equal. Without sustained, consistent, and intentional steps made to ensure an open learning environment, just the presence of a person of Color—even one in charge—does not change the ethos of the classroom or prevent or negate negative racial experiences. Black women and women of Color are often left to fend for themselves, deciding when to obey the rules of racial engagement and when to challenge them, even with a whisper.

Valerie, a Black student in New York, speaks more to the decision to not respond to racism in the classroom.

This one White girl one time says something low key racist in my class and I was like, "let me get this A in this class though, 'cause we need to like work together on like a group assignment." . . . Unfortunately, a lot of times like people of Color are learning to push past things just so they can, get their education 'cause that's why we're here.

Valerie knows the risks associated with breaking the rules, and usually chooses to abide by them to protect her status and success as a student. Outside of the college context, many people feel the same about experiences with racism in the workplace. Most of us need our jobs in order to pay the bills, and following the fourth rule of racial discourse, making sure the White folks around us are comfortable, can be protective. But Valerie also discusses moments she chooses to break the rules, saying, "Sometimes I'm like itching for a fight. I'm, like, somebody say something racist right now so I can school everybody in this room." In these instances, Valerie chooses her battleground and is willing to accept the consequences.

Peyton, a Black student in New York, has made similar decisions, saying, "Sometimes I end up being like the angry Black woman. But I prefer well-informed, passionate Black woman." For people of Color, silence in the face of racism can be a burden on the spirit. At the same time, speaking out against racism can be damning. Whether or not folks of Color speak out against hurtful comments, the stakes are high!

OFFLINE NORMS CATALYZE ONLINE NORMS

The people interviewed here believe that while White students are protected by silence, students of Color are isolated by it. Staying silent around issues of race not only allows problematic racial discussions to go unquestioned, but also demonstrates to students of Color that belonging on campus may be contingent on the suppression of their racial identities in certain contexts.

As students of Color silence themselves and smile through uncomfortable interactions, they seem to be mirroring the insincerity they perceive from their White peers when talking about race, even though for them the stakes are utterly different than for their White peers. White insincerity, communicated through niceties, might serve as a vehicle for protecting White people's moral self-image, whereas for folks of Color pretending to be unbothered by problematic racial dynamics may be a matter of maintaining actual safety and self-preservation.

The stifling nature of the rules limiting racial discourse on campus may give students an incentive to turn to the internet as a social space more amenable to talking about race. As we see in the next chapters, the internet structures a distinct style of racial messaging that alters the dynamics behind each of the rules discussed here. The first rule is to not talk about race, but online racial discussions are much more common. The second rule is to observe the niceties, or social norms, of talking about race that can protect Whiteness from scrutiny and reduce the likelihood that folks might be called racist. But as we see in the next chapter, many White folks are less cautious about using PC language in some online settings, which means folks of Color believe they are getting the truth, not the alibi. The fourth rule is that folks of Color need to preserve White feelings and esteem, but in many online spaces this perceived need is reduced, something we explore in chapter 5. As people of Color feel less need to protect White feelings online, this also negates the third rule, because acts of online resistance reduce the extent to which White folks are able to break the rules of racial engagement without consequence, lessening the effects of political correctness and Color blindness on the way students talk about race.

4

MASK OFF

Revelations and New Realities

> DRIVER: "That ain't no way to board a train you damn, stupid ni—"
> BANG! Trudy pulls a SIXGUN out and shoots the driver in the head.
> CHEROKEE BILL: "You know, he might coulda said nincompoop."
> TRUDY SMITH: "We ain't no nincompoop."
> —*The Harder They Fall*, 2021

ONE OF THE RESULTS of the shift from open and explicit racism to hidden, subconscious, and coded racism is that speech about race is minimized in many social settings. White people want to avoid being called racist, and the idea that one does not "see race" has become the preeminent defensive strategy; if one does not acknowledge racial difference, how can one be racist? Mica Pollock uses the term *colormute* to describe settings where individuals are unwilling to discuss race even when dealing with phenomena that are clearly racialized in nature.[1] In the previous chapter we saw how silence about race, or PC language, can safeguard folks from being labeled a racist.

But the self-proclaimed anti-PC crowd, especially including the Alt-Right, has taught us that not everyone is interested in abiding by the rules of laissez-faire and colorblind racism. Historically, conservative politicians have used coded language to sell racist policies—those that hurt Black people and communities of Color—playing on racial fears without going so far as to be overtly racist.[2] But today there exists a contingent of mainstream conservative politicians who have been emboldened by Trump's success despite (or because of) his brazenly coarse rhetoric around race, as well as his past behaviors demonstrating a lack of concern and even contempt for racial suffering. Given the popularity of this new school of conservatives and their refusal to talk nice about marginalized populations, it is clear that many Americans are annoyed at the idea that they have to pretend to not see race. In fact, research shows that this sentiment is not limited to the far-right; 80 percent of political moderates believe that political correctness is a problem.[3]

Anyone who has ever read comments under a YouTube video or online news article can see that not all racism in today's digital age is subtle. Online racism is ubiquitous. Racist memes, jokes, and pranks are everywhere, and are often more explicit than their face-to-face counterparts. For many internet users, and young people in particular, online racism has become a fact of life. Brendesha Tynes and colleagues found that around 70 percent of surveyed high school students reported witnessing online racism.[4] From the thousands of racial slurs posted on social media each day, to the resurgence of White nationalism through a rebranding as the Alt-Right, to the forty-fifth president's official Twitter account, many online spaces provide opportunities for people to be openly racist without social consequences.[5] Whether it's White sorority members being expelled for posting racist rants on social media on Martin Luther King Day, students being kicked out of school for posting racist-themed pictures on social media, or a White police officer being fired when private social media-based messages encouraging other officers to shoot Black people became public, we can't seem to go a month without an incident of racism online that makes the news.[6]

It is more tragic than ironic that the collective outrage around a police officer *talking* online about killing Black people would lead to that officer's firing, whereas police officers who actually kill unarmed Black folks go free so often.[7] Perhaps this can be explained by the visibility of racist intent on the part of the former, who displayed an explicitly racist attitude in his online messages. His posts were like an admission of racist intent, which, as we discuss in chapter 2, is the ultimate racist litmus test for many in America. In contrast, racial bias in policing is still largely masked by seemingly race-neutral and subjective decisions around what constitutes "reasonable suspicion" or perceived threats to officer safety. It is the overt nature of online racial incidents that engenders surprise and outrage. These incidents are reminiscent of Jim Crow–era racism, a monster many Americans thought was destroyed with the civil rights movement.

This is racism *unmasked*. Whereas we have understood post–Jim Crow racism to be masked behind friendly interactions and supposedly colorblind laws, it is being unmasked in online spaces that reveal the types of open and ugly racist attitudes and actions that many people thought were behind us.

Research has given us a good sense of how ugly things can be online. But how does this affect people? Is online racism something that young people rationalize away and are unbothered by? Does it have any impact on the way they understand racism in a world that is sometimes said to be colorblind? The ugliness of online racism seems to be in stark contrast with the cautious, politically correct norms of discussing race on college campuses and many workplaces. What happens when these two realities collide?

In this chapter I present a case study of an on-campus event— something that occurred in Chicago—that provides a unique opportunity to answer these questions. The campus had been rocked by the appearance of a controversial Facebook page titled, "Politically

Incorrect Confessions" (hereafter PIC). The page was purportedly designed to give students an outlet through which they could anonymously share their honest opinions about race, religion, gender, sexuality, and any number of other topics that they may have felt uncomfortable discussing in person. In practice, however, the site became a safe haven for hate speech.

Over a period of eighteen months, nearly nine hundred comments were posted to the PIC page. The technological setup of the PIC page allowed students to anonymously post explicitly racist language without needing to worry how they would be perceived by their peers. Anonymity can complicate research design because there is no way to tell who is behind anonymous comments on a news site or interest-based web forum, which limits the conclusions we can draw from anonymous online data. The structure and content of the PIC page, however, provides more clues about some group identities of the anonymous posters than do anonymous online comments in many other technological contexts.

For example, users were instructed to submit "university-related" content to the page moderators, ensuring that while comments were made anonymously, they originated from a specific educational community. Many PIC comments made specific references to settings, groups, and individuals on campus in Chicago, so it was clear that the users behind the racist comments were indeed members of the university community. This made the page even more troubling than anonymous racism in other online contexts, which is just as ugly, but less targeted, and rarely coming from within a given community.

By combining posts from PIC and interviews with students, I am able to tell a more complete story than I could with just one source of data. I talked with students about the PIC page and how it affected the way they understood racism and race on campus and in the world. For many students, it represented the first or strongest connection between racism they experienced online and their day-to-day lives in the real world.

POLITICALLY INCORRECT CONFESSIONS

On the front page of the Confessions Page is a picture of a construction-style sign reading:

WARNING: POLITICALLY INCORRECT AREA: ALL P.C. PERSONNEL ENTERING THESE PREMISES WILL ENCOUNTER GRAVELY OFFENSIVE BEHAVIOR AND OPINIONS: RAMPANT INSENSITIVITY AUTHORIZED.

Below the sign were instructions for anonymous posters saying:

Share your [Chicago] related, politically incorrect thoughts and feelings. We'll say them for you, so you don't have to!

Through a submission link on the PIC page, Facebook users could send site moderators their comments, which were then anonymously posted to the page. Posts were numbered so that users could reference previous posts in their comments. And while original posts to the site were anonymous, if Facebook users wanted to reply directly to a comment their profile names would be visible.

André Brock writes about the importance of analyzing how an online interface reflects the intentions of the interface's creators and shapes the creation and interpretation of the interface's messages.[8] Applied here, it is clear that the PIC was designed to facilitate a particular type of posting. For example, by making the comments anonymous, moderators provided a way for students to separate their PIC posts from their real-world identities. This feature was hugely important for some users, as one post read.

I think it's a positive thing that this offers a forum for people to voice some of their most controversial thoughts without fear of character assassination. There are many lived realities on the surface of this planet. Using the type of safe, institutional language and rhetoric favored by the Provost or the US State Department does not always do justice to these realities.

Occasionally, even the language and examples which we feel comfortable publishing under our own names do not suffice to articulate a truth.

This poster does not detect the irony in lamenting the need to use safe language on campus to make others feel comfortable, while simultaneously celebrating the increased safety and comfort the PIC provides for students who want to post offensive content. The PIC creates a safe space for a subset of students who are opposed to the existence of safe spaces for those with marginalized identities on campus. These students are concerned about "character assassination," a point that is further illustrated by another poster.

This page suggests to me that [Chicago] students struggle to navigate a society extremely focused on being politically correct. In my mind the level of political correctness has reached the point of stifling discussion and debate on issues of race, gender, sexuality, etc. I can't have legitimate discussions with anyone outside of my immediate friend group without coming across as a racist.

One form "character assassination" can take, it seems, is being labeled a racist. In the modern world where we have *racism without racists*, the term *racist* is often colloquially reserved for old-fashioned, overt racism.[9] To be labeled racist is to be seen as being morally inferior, uneducated, and backwards. As we discuss in chapter 3, many White students seek to distance themselves from this characterization.

In chapter 2, I discuss Picca and Feagin's two-faced racism theory, which differentiates between the way whites talk about race in public, the frontstage, and in private, the backstage.[10] Here the poster talks about the difference between their frontstage and backstage conversations. They intentionally limit racial discourse to the backstage to avoid engaging the "stifling" frontstage campus atmosphere. The PIC page, however, operates in a hybrid space that functions as both frontstage and backstage; the poster can publicly share the type of thoughts typically reserved for backstage interactions, without facing the consequences that are associated with those types of thoughts being shared frontstage.[11]

In this example, the poster does not believe that their conversations about race in the backstage are in fact racist. Instead, so the thinking goes, this is a label unfairly applied to legitimate intellectual positions. Not all users agree, however, that what the PIC protects is simply a matter of semantics. For example, another user has an even more controversial understanding of the PIC's virtues.

I hate when people argue against hurtful speech without actually knowing why. People presuppose that racism is bad and political correctness is good without even challenging their own assumptions. Furthermore, I hate how people are criticized for defending this page and defending freedom of speech.

Unlike the previous two posts, which both suggested that the racist label could be unfairly applied to controversial but defensible ideas, here the poster suggests that racist ideas are sometimes legitimate social positions. This post calls into question the idea that the PIC was created to foster dialogue. Instead, for this poster, the PIC page represents a protected social space where taboo racist ideologies can be communicated without condemnation.

While not all campuses have their own confessions page, the perception that there are students who feel unable to share their true feelings about race is widespread. For example, Judith, a Latinx student in New York, talks about a similar sentiment that she perceives on her campus, saying, "And then also there's been like, the silent majority within the [university] that is very right-leaning . . . That's how they would refer to themselves . . . they see themselves as having their voices silenced because [the university] is, like, such a left-leaning student body." It is possible that the feeling of being silenced, the desire to voice one's true views about race, permeated all four campuses, and indeed research has explored similar anonymous messages on other campuses through the YikYak app.[12] But Chicago is the only place during the course of my study where technology provided opportunity for these views to be shared without social or educational consequences.

Not all of the posts on the Confessions Page were race-related or offensive. Some posts were innocuous or personal, concerning student crushes, confessions about mental health, or academic struggles. But when I asked students of Color about the PIC page, it was universally remembered as a social space dominated by racist, sexist, and homophobic content. These examples from the page illustrate the tone of racial messaging that was frequently seen on the site.

> Every time I see a black person with an iPhone, I reach into my pocket to make sure it isn't mine he's holding.
>
> I honestly love going to [a neighborhood fried chicken restaurant] and eating the fried chicken special, some watermelon, and grape kool-aid. Makes me feel like a true brutha from da hood.
>
> As if we need any more proof that Asians can't drive whether it's a car or an airplane. Maybe those pilots should turn to another profession, like cooking. After all, they really outdid themselves with that Korean barbeque à la Boeing today.[13]
>
> I'm predicting that the top three items sold at the upcoming Whole Foods store in [a Black neighborhood near campus] will be organic fried chicken, organic watermelon, and organic grape drink to wash it all down. The real question is, will they accept [welfare] cards and food stamps?

In each of these posts we see the expression of overt racial stereotypes. This is not the type of subtle racism college students typically are exposed to on college campuses. These posts also appear to be attempts at irony or humor. While research shows that White men in college make racial jokes often, they are typically careful to avoid making these jokes in front of students of Color.[14] The PIC page thus exposes students of Color to the types of racial jokes, comments, and stereotypes that are typically reserved for backstage, private interactions. But while these posts stand out from the way race is talked about in-person, they are consistent with the literature on race-related com-munication in anonymous spaces on the internet.[15] Across the board, participants from every university talked about the racism they saw as normal online, from YouTube Comments to social media.

What distinguishes the PIC page from the types of racial messaging that typifies some anonymous online spaces, however, is its connection to the Chicago community. For example, if a student reads racist com-ments on a Fox News article, these comments may be hurtful, but they are also more distant. The posts could have been made by anyone, any-where in the world. Students may make assumptions about who the anonymous posters are, and may assume they're the exact opposite of who they go to school with. These assumptions about anonymous rac-ism on the internet cannot be made about PIC posts with similar con-tent, however, because they are made by Chicago students, about Chicago students. For example, another post reads:

> Seriously WTF is up with the onslaught of black people invading [Chi-cago]? My first year, there were barely any on campus, and now they're infiltrating the [library] too (of all places.) Now, I'm not usually racist but they really need to stop being so fucking LOUD. Seriously, if I wanted to be surrounded by a cacophony of vernacular I would've taken the [train] down to [a Black neighborhood].

"Onslaught," "invading," and "infiltrating;" this language signifies that Black students are perceived to be part of an unwelcome and danger-ous group of outsiders whose presence is having a forced, negative impact on the Chicago campus community. In fact, it mirrors the type of anti-immigrant language associated with the Latino threat narrative we talk about in chapter 2.

The campus is near African American communities and students receive racially coded formal and informal messages around safety on campus that symbolically separate the safe campus space from its sur-rounding community, which is depicted as violent and dangerous.[16] For example, students talked about being warned away from visiting Black neighborhoods on campus during their orientation, and received regular emails detailing crimes committed on or near campus, often with an ambiguous description of the suspect that could refer to many Black and Latinx men on campus. This PIC post suggests that Black students are perceived to have more in common with the dangerous other that represents the university's surrounding communities than students on campus.

This is not unique to Chicago, of course. When I was a professor at Boston University, I received an email about a suspect on campus who had a "dark complexion," "distinctive orange backpack," and was riding a bike. I saw this email on my phone while I was getting on my bike, wearing my distinctive burnt orange backpack. Instead of riding my bike home with my bag and laptop to do some work in the evening, I left my bag and bike in the office and walked home. After more than a decade in academia, I was used to feeling that I fit the racialized descriptions of suspects in campus safety emails. But this email was referring to a suspect wanted for a violent crime, and I thought it could be dangerous to fit the description so closely, down to my accessories and mode of travel.

Many experiences with online racism are indirect. But this post is different. It appears to be written by a student in Chicago, targeting Black students in Chicago by referencing a specific interaction. This type of targeting may increase the harm caused when compared with the "random" racist posts that appear often in various online spaces. Imagine you're a Black student who has occasionally laughed in the library with your friends (as students of all races do regularly). Will you question whether this post was written about you, specifically? Would that make you less comfortable using the library? When I was an undergrad, the library was the only place I could get work done. The temptation to socialize or play video games was too high anyplace else. I'm not sure what my grades would have looked like had I received an anonymous message telling me that Black folk were unwelcome in the library.

Another post also speaks directly to the marginal social positioning of students of Color in Chicago.

> It's odd to note that minorities made up a higher percentage of students in the Intro Chem section in comparison to the General and Honors sections.

By prefacing the statement with "it's odd," the poster frames this discussion as a legitimate question that is guided by curiosity, not racism. Nevertheless, by suggesting students of Color are concentrated in low-level classes the post questions whether students of Color are smart enough to handle academic rigor at the university. Research shows that when Black and Latinx folks feel like their deservingness is being questioned—for example, if it is suggested that they were only admitted because of affirmative action—they will be less likely to seek help or access campus resources that are meant for all students, which increases the chances of harmful stereotypes becoming self-fulfilling prophecies.[17]

Here we see that the PIC is a context that not only provides users with more opportunity to express explicit racist ideologies, but it is also a space where subtle racist ideologies are unmasked. Imagine: narratives couched in notions of individual deservingness are typically representative of subtle forms of discrimination, but in this example the line between subtle and explicit racism is blurred. In the context of the PIC page, subtle racism forfeits its subtlety. Because the post was made anonymously on a site that protects overt racists, it can be assumed that the poster knew that what they were saying was offensive.

RACISM REVEALED

The campus in Chicago was in an uproar when the PIC page dropped. People couldn't believe what they were seeing. For many students, the page forced them to confront the reality that racism was not as dead, or limited, as they had previously thought. Chris is a Latinx student in the sciences who is involved in several academic student organizations. He didn't think race was a major problem on campus, but the PIC page called some of his assumptions into question.

> Before the PIC, like I said there were a few incidents that I mentioned earlier, but I think the overall tone wasn't too bad. Like, I had, you know, good friends, I had good spaces, so it didn't seem that real . . . So then when the PIC page came up, I just realized there are people who hold these things. Maybe they don't say them directly to people, or maybe they don't take action.

The "few incidents" Chris references here were highly publicized race-related incidents involving fraternities, racial epithets, and racist-themed parties. He chose to not define the campus by those unfortunate events, however, instead seeing them as being isolated incidents that were not common enough to reflect his day-to-day experiences on campus. Posts on the PIC, on the other hand, seemed to indicate that people's friendly actions and speech did not always match up with their privately held beliefs.

Why did it take posts on the internet to convince Chris that racism was a problem on campus? It's actually common for people to have a hard time recognizing discrimination when they do not believe it affects them personally. In my classes on race, I have students fill out a "personal privilege profile" (an activity I inherited from my teaching mentor, Dawn Belkin-Martinez), and list whether they have advantaged or disadvantaged statuses on a range of identities, including race, gender, sexual orientation, age, religion, and physical ability.[18] Most students have an easy time identifying their nondominant identities, but tend to be surprised by some of the dominant identities that they do not need to think about on a daily basis. For example, if we are able to walk upstairs without pain or assistance, we may not think about the privilege of accessibility and whether buildings or restaurants have been intentionally designed to accommodate disabled populations. It's harder to think about the ways we might be privileged than the ways we are disadvantaged. If someone describes a social problem that we don't see in our everyday life, we may choose to not believe them. Some people need to see the evidence. For others, no evidence is convincing enough to contradict their personal experiences.

This is something I see even in my own personal circle. I have some Black friends who believe that the difference between them and Black people who are less successful is how hard they work, or their decisions to value middle class, Christian norms over values they would see as being driven by the media or "street" culture. They do believe racism exists, because they experience some form of racism in the workplace, but they are less convinced of how racism is structural and affects the opportunities and limitations people face, things that prevent less privileged Black folks from reaching the same places they have. This is one of the difficulties of teaching people about racism. When you experience the world as good and safe, you may not understand that there are other people who do not experience any of those things.

An example of this this comes from basketball legend Michael Jordan, who has long been criticized for his perceived lack of support of social causes in the Black community (including when he famously said, "Republicans buy shoes too," a statement that was allegedly connected to his refusal to support a Black Democrat running against an openly racist Republican senator in his home state of North Carolina).[19] When I first heard that Jordan put out a statement about police violence in 2016, I was excited, hoping that he would finally begin to speak out against racism, the way athletes such as Lebron James, Colin Kaepernick, or Craig Hodges had been.[20] Instead, he wrote a statement that blamed "divisive language and racial tensions," not racism, and said, "Over the past three decades I have seen up close the dedication of the law enforcement officers who protect me and my family. . . . I also recognize that for many people of color their experiences with law enforcement have been different than mine."[21]

Then, in true *put your money where your mouth is* fashion, he proceeded to announce $1 million donations to *both* Black Lives Matter and a community policing organization. This playing of both sides did not feel like support for Black Lives Matter. Even when he decided to speak out against police violence, he refused to condemn the institution responsible for this violence. While Jordan acknowledged in his statement that his privileged experience was not universal, his actions showed that his positive experiences with police made it difficult for him to understand how the police institution was to blame, despite the overwhelming evidence, both anecdotal and statistical, of racial bias in policing and anti-Black police violence.[22]

Chuck, a Latinx student who is involved in several cultural organizations and is also a member of a majority-White fraternity, gives another example. He has a small group of racially diverse friends and, like Chris, talked about having only positive interracial experiences on campus. He says about Chicago, "Race shouldn't matter . . . we're all academics here." This idea, that intellectual elites are beyond the race problem, was called into question by the PIC.

[They] immediately just started posting just like vile, horrendously racist things and there was like a lot of people that were going like, "Oh yea, like I totally agree with that" . . . At [that] point it's, like, these are . . . these are my classmates and this is like how they feel about like minority groups . . . this is something like that you would read about that like the Klan is doing in the 50s, not like now.

Here we see Chuck forced to confront the incongruence between his perceptions of a postracial campus and the overt racist messages on the PIC, which revealed that friendly campus interactions may belie hidden racist attitudes and challenged his understanding of how racism operates. Some people need the vile, horrible things to wake them up to ugly realities of the world that lie beneath the surface. Actress Jennifer Lawrence talked about how she was a Republican until Trump. While Trump's policies weren't very different from other conservatives before him, his language around race and gender was certainly more offensive. The PIC, like Trump, was like a splash of cold water to the faces of those who were unconscious or unaware of the prevalence of masked racism.

One of the more alarming effects of the PIC website was how it changed the ways students of Color related to their White peers or even Chicago as an institution. Sara was bothered by not knowing who among her peers were responsible for the hateful content online and noted the impact the PIC had on her interracial relationships, saying, "It makes it difficult for me to form friendships with people of different races . . . I know how you really feel." Sara's perception assumes that the ideas on the PIC are widespread on campus—something that may not be true, given what we know about how small groups can have a large voice on the internet. Regardless of how widespread the ideas were in reality, however, the unmasking of racism on the PIC was perceived to reveal the true racial ideologies behind the façade of postracial colorblind interactions on campus.

Lisa, a student from Chicago who identifies as multiracial (Black and White), speaks more to this.

> I know a lot of students didn't necessarily feel afraid, but they definitely felt more like they should be more wary of who they're hanging out with on campus or who their friend circle is or—because they don't know if any of these bigot posts are coming from people in their classes or people they're interacting with.

Lisa's words illustrate the trauma that was associated with this page. For some students of Color who were happy on campus—or at least did not feel like they were under racial attack—the PIC was like discovering they were actually trapped behind enemy lines. This revelation gave them cause to reconsider many of the relationships they used to take for granted.

This feeling held even for those students who talked about being very aware of racism on campus. Rachel said:

> It was so cowardly, because I mean, if you just have raging hatred or igno-rance inside of you, but you wanna own it, then go ahead! You know? Like that's your life. But, for you to wanna play both sides and be secretly racist or homophobic, but then you still want me to smile at you when I see you [on campus], or you still want me to sit by you without walking away, or you still want to be welcomed into smaller circles of the campus commu-nity that are more [vulnerable]. Like, it's just, it was unbelievable.

Elsewhere in the interview, Rachel talked about using a form of informal peer classification before the PIC, identifying and avoiding White students she thought were ignorant or antagonistic around issues of race. If students were open with their "raging hatred," then Rachel's adaptive coping response—coding certain students as "racist" and avoiding them—would work just fine. But the anonymity of the PIC made this impossible. Rachel lost her primary strategy for dealing with racism: the identification and avoidance of perceived perpetrators, even those engaged in subtle, masked racist behaviors. Rachel felt capa-ble of finding and avoiding masked racism on campus, but in this instance the unmasking of racism made it more challenging for her to identify racists whom she notes want to "play both sides" by posting racist comments on the PIC, but still being friendly with students of Color on an individual level. There is no way of knowing how effective Rachel's methods of identifying racists actually were. But for Rachel, the anonymous messages on the PIC reshaped her face-to-face reality of racism by taking away her agency and ability to decide not to inter-act with those she considered racist.

Daveena, a Latinx student in Chicago, said about the site, "You asked earlier if I felt that I belonged at the university, and after the PIC I really felt that I didn't. That I was in a place that was unsafe for me." Daveena was no stranger to racism before the PIC page, and felt confident in her abilities to fight racism through activism. Daveena's work as an activist was clearly defined when she knew who and what she was up against. But because the antagonists on the PIC page were anonymous, Daveena began to see herself as a perpetual victim with a seemingly infinite pool of enemies against whom she was unable to protect herself. Daveena felt distanced from her peers on campus and undervalued by Chicago as an institution. There was nothing she could do to remove offensive posts or confront the anonymous students behind them, and she blamed the university for not doing more to protect her and others.

When some students of Color pressed the school to respond to the PIC, administrators claimed to have limited power because they could not control an external website. The only concrete administrative response to the PIC was to demand that the moderators take Chicago out of the page name—and this did not take place until five weeks after the page first appeared. The makers of the site replaced the Chicago name with the school colors in the title of the PIC, thereby still clearly marking its connection to the university. Jerry discusses why he believes Chicago did too little, too late.

> I stopped doing my work. I was just really in awe that this was something that I was reading . . . it makes you feel really unsafe when you know that there are a lot of students around you who have these very fucked up feelings about you based on what you look like . . . how are we supposed to build community when you're allowing your students to do things that [create] distrust in your community?

The truth is colleges can't actually moderate spaces on the internet beyond their own websites. This is a continuous challenge for universities whose campus climates are affected by things that are posted online but are outside of their "jurisdiction." Harvard University rescinded admission for students who posted racist and homophobic content online, but these students had their names attached to their posts.[23] While Chicago could not have done the same, students needed more of a response from the administration to confirm that students of Color were valued by the university. Hearing from Jerry about how this affected his academic engagement, we see another potential pathway through which the PIC could exert tangible influence on educational outcomes among students of Color.

Not everyone was surprised by the page. Some saw it as proof of the continued significance of racism, something they felt was not universally acknowledged. Gina, a Black student in Chicago, is a student leader in campus diversity-focused and cultural organizations. Her reflections highlight a different adaptive strategy for making sense of unmasked racism on the PIC.

> Racism is nowhere non-existent . . . You could be talking to, like, a White person that you think is your friend . . . that could be one of the people who are really upholding White supremacy on that page . . . I think it's good that people still know that racism is out there because a lot of people just think it's dead and that's, like, not even close to being true.

Gina identified a silver lining: the page is a clear indication that society is not postracial or colorblind. The PIC challenged the racial narrative on campus, making it clear that Chicago, despite its elite, liberal, and "regular" student body, was not beyond the race problem.

David is a student in Chicago double majoring in the social sciences and race studies. Before the interview he said that he was heavily invested in having racial discussions, and knew he was interested in being interviewed when he heard the words "race" and "research" together when I presented the study to the Black club on campus while recruiting students. David is mixed, with Black and Latinx heritage, identifies strongly with both groups, and is involved in both Black and Latinx student organizations. When asked about the PIC, David said:

I actually saw [it] as more of a body of evidence to present to other people to say, "Hey look, all that stuff I was complaining about and talking about. This is it right here in the flesh for you to see, or online for you to see, instead of you thinking I am crazy coming up with conspiracy theories. This is it right here."

Throughout our interview, it was clear that David understood racism on a deep level, putting his experiences on campus into the broader context of White supremacy, and that he was familiar with ideas like White privilege, structural disadvantage, and covert racism. So, despite the general absence of presentations of traditional racist ideologies on campus, he was not surprised by the attitudes expressed on the PIC page, even though he was taken aback by their open presentation. About one-third of the people I talked to echoed David's conflicted mindset: being simultaneously surprised and not surprised by the site. Surprised because the PIC was contrary to the way racial discourses typically played out on campus, but also not *really* surprised because they knew that racism existed, even if it was being masked by gentle racial discourses.

For students like David and Gina, the PIC was like a distasteful video of police violence against Black people being shared on the internet. It can be traumatic to see those videos, and many Black people would prefer to not have videos and images of Black death and pain on their social media feeds. Yet those videos are interpreted by many as proof of the continuing realities of racism. Online videos that are leaked or go viral can be used as evidence that racism and racist violence are normal, everyday occurrences that cops typically get away with because there is (typically) no proof, and that are only being talked about because of the video. How many more people have been hurt by police off camera, than have been witnessed being hurt by police on camera?

This brings us back to the puzzle from the first chapter: what do online expressions of racism tell us about racism in the real world? For some people, the answer is *everything*. Cristina, a South Asian student in Chicago, said, "In some ways it kind of took the trigger [of] something as appalling as PIC for people to kind of be like, "Oh wait, we think racism isn't a thing on this campus. Turns out, it totally is . . . Let's talk about it."

In a paper articulating the linguistic style of colorblind racism, Bonilla-Silva suggests that language that can sound racist is seen as taboo, and that "because the dominant racial ideology portends to be color blind, there is little space for social sanctioned speech about race-related matters."[24] But Cristina suggests that the PIC shattered notions of colorblindness amongst students in Chicago, weakened the color-blind rationale for why race-related matters did not need to be discussed, and led to an increase in the number of conversations being had about race on campus. The PIC challenged the perceived dominance of colorblind discourse on campus, and for many students that was an indication of how problematic the racial climate in Chicago was, despite the rarity of overtly racist incidents on campus. This led to an increase in online and in-person discussions of race that students of Color suggested had been easier for Whites to avoid before the PIC.

SEEING UNDER THE HOOD

The technological mechanisms through which the PIC rewrote the rules of racial discourse, as well as the processes through which this discourse influenced the ways students of Color perceived the campus racial climate, demonstrate the potential impact of technology on the way we think about and experience race and racism in the twenty-first century.

Examples from the PIC enhance our understanding of post–Jim Crow racist ideologies. Posters on the PIC reported feeling pressured to act according to dominant colorblind styles on campus and resented the perceived need to censor their thoughts on race and mask their true racial ideologies. Internet-based communication, like the PIC, can provide opportunities for individuals to bypass the surface-level egalitarian norms that characterize many mainstream spaces, thereby enabling more explicit expressions of racist ideologies.

It makes sense that students of Color perceive direct connections between online racial messages and campus racial climate, interracial relationships, or institutional support. Many colleges and universities wrestle with the sometimes competing ideals of free speech on campus and the desire to protect students from harassment, and some schools have punished students for their involvement in circulating racially or sexually offensive content online.[25] As our understanding of institutional climate expands to include online social spaces, schools must develop policies formalizing approaches to moderating these spaces or responding to student concerns about online racial messages that impact campus climate or student well-being.[26]

Students who had long been able to articulate the subtle nature of contemporary racism were able to use examples of unmasked racism from the PIC as a vehicle through which they could help others identify the manifestations of racism. Given the dominance of masked racism in mainstream institutions and interactions, this is one way the online unmasking process can expose racism, both on and off college campuses: it highlights the embeddedness of racism in interactions and institutions that, prior to the unmasking, were only visible to those with the language to understand the way racism operates in seemingly race-neutral interactions or institutional practices.

For example, while calling the police is not commonly seen as being a racist act, in the past few years social media has driven a renegotiation of popular conceptions of racism through the application of the "racist" label to Whites that call the police on Blacks for no apparent reason.[27] Social media users have thus used technology—sharing and commenting on videos of White people calling the police on Black people as they wait in Starbucks, barbeque in public parks, or go swimming—to highlight the way Whites have the ability to wield a public institution to harass and potentially endanger Blacks and People of Color.[28] This represents an unmasking of racism at interpersonal and structural levels; it problematizes not only individual-level prejudices that led the Whites in the videos to assume criminality and call the police, but also the ways purportedly race-neutral institutions can reproduce and reinforce racial inequality. Apryl Williams writes about these memes as increasing agency for Black folks and subverting racial power structures by demanding consequences for racist actions that had previously been framed as being innocent.[29] White privilege works best when it is goes unnoticed and unchallenged. When folks use technology to challenge privilege, they help others better understand how it works.

Racism is unmasked in online spaces, therefore, not only when internet-based communication facilitates the sharing of overtly racist sentiments, revealing these old-fashioned racist ideologies to be alive and not defunct, but also when online users highlight the previously hidden mechanisms through which contemporary covert racist structures are maintained. Just as unmasked racism on the PIC changed the way students in this sample thought about race and racism in Chicago, the unmasking of racism in other online spaces can challenge dominant racial ideologies and expose more people to the ways racism works in the post–Jim Crow era.

CODA

In late 2021, a video of Judge Michelle Odinet repeatedly yelling the n-word went viral. After being placed on unpaid leave, Judge Odinet resigned from her position. In response to the outcry, District Attorney Jason Williams launched an investigation into Odinet's cases as a prosecutor in New Orleans in the 1990s, looking for evidence of racial bias.

Judges don't need to use racial slurs and call Black people cockroaches, as Odinet did, in order to give Black people longer sentences or charge them exorbitant court fees, whether they are "innocent" or "guilty." The response to Odinet is another example of how many people respond to open, obvious racism differently than masked, invisible racism that is embedded in systems and institutions, from the criminal justice system to school systems to the workplace.

The evidence of racism with a capital *R*—a video of Odinet using overtly racist language which has been shared and viewed millions of times on the internet—was enough to force a second look at her supposedly impartial judicial decisions. Other judges may have said worse, but been crafty enough to not be recorded on video, and won't have their cases investigated for evidence of racial bias. Still more judges may never use the n-word, but if we were to open their books, we'd find that they are more racially biased than Odinet in their courtroom-based decisions.

Structural racism, more than interpersonal racism, is the primary mode of racism that hurts the lives and life chances of Black people and people of Color. As we've discussed, structural racism does not require overt bias to function. For me, looking at data that shows the correlation between race, poverty, and poor performance of schools makes me as angry, and is as strong of evidence of the existence of racism as video of White students using the n-word. But because evidence of structural racism is less flashy, it can be less convincing for some people. The unmasking of racism in online spaces is so important because it makes racism visible to people who cannot recognize masked racism. And as we see in the next chapter, this increased recognition of racism may increase the number of people who are willing to resist.

5

DIGITAL RESISTANCE

You cannot tell a [hu]man they have the power to make the earth
shake, then expect them to walk small.[1]
—ROBERT JORDAN, *Lord of Chaos*, 1994

DO YOU HAVE AN ID?

As a PhD student at the University of Chicago, I had 24-hour
access to the computer lab. I would often go home at 3 or 4 in
the afternoon, spend time with my kids, and then come back
in the late evenings to do work after they went to sleep.
Many nights, if I worked until 2 or 3 a.m., the only other person in the building would be Gary, the nighttime custodian.
Gary would stop by the PhD lab and chat, and he and I developed a rapport over the years.

The school was open to the public until 9 p.m. After 9,
however, we had to swipe our school IDs on the door to
open it, like a hotel key card.[2] One evening, I arrived at the
school at about 8:55pm for a nighttime work session wearing
a hoodie. I had my school ID with me, but I didn't need to
use it to swipe in, as it was still before 9pm. A tall White
man wearing a blazer, layered over his sweater and dress

shirt, was walking out of the door as I was coming in. Instead of walking out, however, he stood in the doorway, blocking my entry. "You have to have an ID to get in here," he said. While he clearly didn't recognize me as a doctoral student, I recognized him, an esteemed professor.

This request caught me off guard. It was too early to need an ID to get inside. Does he ask everyone for ID? If not, is he asking me for an ID because of how I am dressed? Because I'm Black? He's blocking my way—is he ready to try and physically stop me from entering the building? I wanted to ask him these questions, but knew that if I did I would risk offending a powerful professor, something that could derail my young academic career. Should I then just show him my ID? While I knew that this would be the "safest" response, I also knew that I would regret compliance with what I perceived to be an unjust demand. After all, I thought, I had the right to be in this space as much as he, or any other PhD student—and I was fairly certain that he wouldn't be blocking my way into the building if I were White.

So I decided to respond in a way that wouldn't accuse him of any wrongdoing, but also allowed me to deftly stand up for myself. I slowly reached for my wallet, taking it out of my back pocket while saying, "Oh, are you the security guard?"

"No, I'm faculty," he said, poking his chest out.

"Oh, ok," I said, putting my wallet back in my pocket without opening it, and walking past the professor without looking back. While I was scared to confront him openly, I communicated what I wanted to without words. If a campus police officer asks for my ID, I am obligated to show it. But if you are not an officer or security guard, you have no business asking me for my ID, let alone physically stopping me from going anywhere on campus.

I think he got my point, because he yelled after me, somewhat defensively, "I'm just saying, at night you have to show your ID—the maintenance crew will ask you for it."

As fortune had it, Gary was just a few feet in front of me, doing some work in the lobby. "What's up, Gary!" I said loudly, dapping him up[3] like I always did.

"Oh, yea, he's good doc!" Gary said to the Professor.

As I walked away from Gary and the lobby, the professor said something else that I don't recall. I didn't turn around or slow down, I just walked down the stairs and made my way to the computer lounge.

This experience, though it seems minor, is something that I have reflected on for years. It impacted my experience in grad school—who I made connections with, and the trust I felt for the institution. And unfortunately, it wasn't unique. Besides not being the last time I was carded on campus—the next time was while I was playing basketball in the university gym (suspicious, I know)—it is also something experienced by many people of Color on college campuses.

I share this experience because I think it can highlight the ways racism often works in the twenty-first century. Unless you're at a White supremacist rally, overt racism has mostly gone out of style. When the media catches wind of overtly racist statements, all parties involved are quick to give a public apology and make sure everyone knows they, or their organizations, do not adhere to such abhorrent views. Outside of seemingly outlier cases, in daily, mundane life, people don't usually brag about their racism.

Instead, most interpersonal racism that people of Color experience on a day-to-day basis is more subtle, which can make it tricky to identify and respond to.[4] Experiences with subtle racism or microaggressions may leave us with questions: Was that really racism? What should I do about it? What does it matter anyway? Exploring these questions may help us understand why microaggressions are actually very destructive, why responding to racism is so hard and so important, and how technology has the potential to change this dynamic.

Subtle acts of racism are interpersonal manifestations of White supremacy. Remember, racist language has a social function—it legitimates and justifies the racial divide. During Jim Crow, explicitly racist language was needed by White people to maintain the social hierarchy by indicating that Black people were less than human, unworthy of citizenship, and deserving of unequal and brutal treatment.

Today, the line between White folks and folks of Color is not so clearly drawn that this harsh language is needed. Instead, microaggressions fill the same function—legitimating racial inequality—but they do so without harsh, ugly language. For example, while there are no signs on campus saying, "WHITES ONLY," neither are Whites asked to show ID to prove that they belong.

The term *microaggression* was coined by Chester Pierce, a Black psychiatrist who used it to describe the everyday, subtle experiences Black folks have with racism.[5] While this concept originated in academic literature, it has become a buzzword of sorts. Between 2012 and 2019, the usage of the term *microaggression* on Twitter increased by over 8,000 percent.[6]

Derald Wing Sue, whose decades of research on microaggressions has shaped the field, writes that the most common way to respond to microaggressions is to not respond.[7] Sue gives six reasons for this, including: (1) not being sure a microaggression has occurred, (2) the tendency for targets of microaggressions to convince themselves the microaggression did not actually occur, (3) the belief that responding to racial microaggressions is futile and won't change anything, (4) not knowing how to respond, (5) not being able to respond quickly enough, and (6) fear of potential negative consequences for responding to racial microaggressions due to racial power dynamics in various social contexts.

When no one responds to or critiques subtle racism or microaggressions, the racist assumptions behind these acts go unchallenged, and the subtle ways some groups are treated better than others is seen as normal. Silence in response to racism, however subtle, strengthens and reproduces White supremacy.

This chapter is about students of Color making use of technology to change this pattern.

WAS THAT REALLY RACISM?

The professor who stopped me from entering the building did not tell me he did it because I was Black. When he told me the maintenance crew would ask for my ID, he was suggesting that asking for my ID was simply a matter of policy, not personal bias. How then can I be sure that his demand to see my ID was because of my race?

I can't. I can never presume to know for certain another human's motivations or private thoughts. Why, then, did I *perceive* this interaction to be about race? Could I have been wrong? Oftentimes racism, especially subtle, masked racism, is ambiguous. How then can we interpret subtly racist acts with confidence?

One way is to avoid the temptation to think about experiences with racism as isolated incidents, as if they occur in a vacuum with no historical, structural, or organizational context. My experience was not exceptional. It was part of a pattern of how Black people and people of Color are often treated in public spaces. Many Black people I have talked to, including students I interviewed for this study, as well as friends at University of Chicago and other schools, talk about the experience of being "carded" on campus for no reason. For example, my good friend JT, a Black man who graduated from the same PhD program as I did, was asked to show his school ID every time he went into the law school library for weeks (and during this time he never saw anyone else asked for their IDs). He would walk in with a group of White students, and each time was the only one singled out. Eventually he worked up the courage to ask the security guard why they chose to ask him, and only him, for an ID time and time again. After confronting the guard, they stopped asking him for his ID—but there was never any admission that race had something to do with why he kept getting carded. While JT was confident that his being stopped had to do with his race, for some there might still be uncertainty. I also want to point out that in this example, the security guard was Black. Just as we see in chapter 3 that Black people often blame other Black people for their own social problems, people of Color can internalize and act on racist stereotype and prejudices.[8]

For many people, including both folks of Color and White folks, determining whether something is racist comes down to intention. Without an admission of guilt, we may feel the need to search for an alternative explanation—something other than race—that can account for what happened. Do we *know* they are racist? Because we tend to reserve the Racist-with-a-capital-*R* label for the old-school, hateful style of racism that argues people of Color are biologically inferior, accusing someone of racism has a moral implication—"that person is bad," or an ontological implication—"that person is permanently defined by their racism." If being racist means someone is irredeemable corrupt, then people may be loathe to assign the label "racist" to anyone whom they believe to be a nice person or someone who might have good intentions. Progressive-leaning people who see racism as a problem limited to conservatives might hesitate to call anyone who leans left politically, racist. Anything less than the n-word, or an open admission of Racism with a capital *R*, and the jury will be out *Was that really racism?*

Would it be easier to identity subtle, masked racism if we did have the ability to read minds? Sometimes, yes. But on the whole I don't think it would help much. Because as we see in chapter 2, it's completely possible for someone to be well intentioned—to have no hateful thoughts about people from another race—and still engage in racist acts.

Research on racial microaggressions finds that these experiences with racial slights occur according to predictable themes.[9] One of the more common themes that Black folk experience, for example, is assumptions of criminality, as both authority figures and overzealous citizens assume they are dangerous or up to no good. One manifestation of this is the constant questioning of whether Black people belong in a space. Hence the question, *Do you have an ID?*

For other groups of Color, the most common themes are different. Asian Americans and Latinx folks, for example, may be more likely to be asked where they are from. No matter how long they, or their ancestors, have been in the United States, they are forever assumed to be foreigners. Another microaggression related to this assumption is the way that folks from Latinx or Asian backgrounds are complemented on how well they speak English—even when they are native English speakers. Because they are assumed to be foreigners based on their appearance, their speaking English with an American accent can be taken as a surprise.

Using these examples, it's easy to see how a well-intentioned White person can commit a subtle act of racism, or microaggression. Microaggressions can be committed without malice, and well-meaning White people might be horrified to learn that their friend or co-worker was hurt by what they thought was a complement—*You're so well spoken*. But the positive intent behind that "complement" does not undo the pain it causes, nor the way that statement reinforces to the person of Color the notion that they do not fully belong in that space.

So even if I could read that professor's mind, it wouldn't make me feel better. He might not have *thought* the interaction was about race. He might consider himself to be an antiracist, someone working for racial justice—and he might live a lifetime of doing good work in that area that I can respect. But in his interaction with me, whether he intended to or not, he reinforced a dominant, hurtful narrative: that Black men were dangerous and therefore unwelcome on campus at the University of Chicago.

When teaching classes on race, I have received feedback from students of Color who tell me that they have had experiences with race that they didn't recognize or fully understand before taking the class. After reading an article or engaging in a class discussion around subtle or systemic racism, suddenly they see the common denominator and reinterpret why they were followed in the store, why the clerk wouldn't touch their hand, why their friends complement them for being articulate, or seem to always want to touch their hair. Subtle experiences can hint at systemic issues, and structural-level inequities can manifest themselves at interpersonal levels. How many Black students need to be asked for their ID on a campus before this is recognized not just as an interpersonal microaggression but as a systemic or university-level issue—with folks of Color being less welcomed on campus? As an individual experiencing those things, they can be disempowering and confusing, and we may not understand how they fit into the pattern. The more we seek out and accept the narratives and stories of marginalized folks, the more we will understand both our own experiences and the ways subtle systems of oppression operate.[10]

Social media and internet technologies make this easier, enabling the creation of spaces where folks of Color don't feel so alone when discussing their experiences with racism. These technologies have increased access to counternarratives that challenge the dominant, colorblind ways of thinking about race and racism. Some basic characteristics of online communication—such as the ability to interact with others without being in their physical presence, and being able to reflect and engage on your own time—may increase the comfort level many people have talking about race. The result, according to the students I talked to, is that discussions about race are much more common online than they are offline. More people talking about race means that students are exposed to a broader set of ideas. For instance, Dolores, a Black student in Chicago, shares how many of the online racial discussions start.

The friends I associate with they're very like, "activisty" you know? So they're always sharing articles, like, "What do you guys think?" And also, here there's a lot of academic debates if there's anything going on in the world, they really want to talk about it and see other people's point of view.

Students interested in generating discussions about race on campus can be frustrated by the perceived lack of outlets. Sometimes they feel talking about race is limited to what the literature calls counterspaces: the clubs, study halls, organizations, or informal peer groups that we first discuss in chapter 3, spaces for resistance and cultural expression.[11] On the internet, however, "activisty" students, or students who are racially conscious or engaged, have more agency in starting conversations about race. They need not wait until the monthly Black Student Union meeting to talk about the latest Ta-Nehisi Coates piece, or an Angela Davis essay they stumbled across. Instead, by simply sharing a link they can start this discussion immediately. And they may be interacting with a larger group online than they typically would in person. In my dissertation, and a later paper, I use the term *online counterspace* to refer to "permanent or temporary online spaces created by students where critiques of covert racism, like racial microaggressions, are normalized and supported" and where discussions about race are privileged and welcomed.[12]

When folks use technology to demand space for discussing racism, from personal experiences with microaggressions to the machinations of systemic racism more broadly, they create online counterspaces. My use of the counterspace concept comes out of critical research in the sociology of education from scholars such as Tara Yosso, Daniel Solórzano, Andrew Case, and Carla Hunter, but in the communication sciences scholars such as Sarah Jackson, Moya Bailey, Brooke Foucault Welles, and Mark Lamont Hill have used the term counterpublics to refer to similar anti-oppressive practices on Twitter.[13] These terms are related, and it excites me that we see networks of folks of Color using technology to challenge the racial status quo both on campus and off campus.

Bringing these discussions online has important consequences for how race is experienced. Rebecca attends both Black and Latinx student groups and spends a lot of time engaging with race issues. She laments the fact that on campus she only talks about race once a week or less. Online, however, she notes that race is "wildly discussed." When talking about these online discussions, Rebecca says, "It also gives us more room to process. It gives you more room to keep thinking and to find space, because people are so much more willing to talk and also answer questions when asked, like, appropriately."

Rebecca appreciates the way she can take her time to reflect before engaging. I also want to draw attention especially to what it means to "find space" online in a way that she cannot in person. On campus, there are few spaces where race can be discussed. For example, Rachel notes:

> If you would take the same conversations or the same ideas that you discuss online and try to put them in the school cafeteria or the library . . . it definitely wouldn't be as welcoming of an environment. People seem unwilling to listen, and people usually in person like to deny that there is a race issue from what I've experienced.

Not talking about race reinforces colorblind norms and maintains the first rule of racial discourse, as we discuss in chapter 3. In contrast, when Rebecca says she can "find space" online, she is referring to the way "activisty" students create online counterspaces. Finding these spaces is important for all the reasons that in-person counterspaces are important: they affirm racial/ethnic identities and facilitate critical thinking. But unlike on-campus counterspaces, which are typically exclusively populated by folks of Color, online counterspaces are racially integrated. The ability to craft mainstream spaces of resistance, visible to people of all races, gives students of Color increased agency in molding the racial climate of their institution.

On campus, counterspaces are meant to be safe spaces for students of Color, places where they can experience a respite from microaggressions. In contrast, online counterspaces are not safe spaces, a reality that both brings out the ugliest parts of humanity and creates unique opportunities for challenging racism. The same characteristics of online communication that make racial dialogues easier to hold, remember, also make them places in which it is easier to share fringe or offensive ideas. Ahmed, a black student in Chicago, talks about what he sees as the negative and positive aspects of social media-based racial discourse. While the positives include increased discussion, attention, and activism around issues of racial justice, he characterizes the negative as being the way many people "hide behind the veil of social media to pretty much say whatever they want, specifically regarding race."

Similar points were made by a number of our interviewees; many of them were glad for the increased authenticity, even if it meant they had to see more offensive content. Jordana, a Black student in New York, says:

> I think that people can be more—like, less hesitant . . . since you're not getting, like, the face-to-face reaction of someone, you can kind of, let's say, get your whole point across before they kind of have the chance to have a rebuttal. But when you're face to face . . . once a lot of emotion is in it, it's almost—it can't really be a discussion anymore, it's just too charged.

Productive discussions about race can be difficult in contexts where both sides not only experience heightened emotions, but also use different paradigms for understanding race, and therefore talk past one another. For example, as we discuss in chapter 3, the dominant mode of thinking about race in ahistorical and individual terms limits the extent to which some White people (or people of Color who subscribe to the dominant framing of race) can understand how historic or systemic racism affects the lived realities of people of Color today.[14] Having immediate access to sources and data, in this example, gives students of Color the confidence and tools to poke holes in dominant paradigms for understanding race. Rachel shares how increased access to empirical resources and time to craft careful statements impacts the norms around online racial discourse.

I didn't really care about grammar, but if you're on Facebook and you're talking to a [college] person, you should make sure that you have, like, a nice thesis, an intro, some citations . . . it encourages you to sit and kind of meditate on what other people say before you respond and type out a very thoughtful response, and make sure you have evidence to back it up.

This thoughtfulness behind online posts seems counterintuitive, when we think about the often-inflammatory messaging that can take place online. Here Rachel suggests this might be unique to online discussions among college students, but I've seen this trend in other places as well, when I scroll through my social media feed or glance at comments on news articles. When debating issues, users often ask for evidence and dismiss the opinions of other users who refuse to share any. Other users post links. This perceived need for evidence explains why students feel it is important that they have time to gather their thoughts and carefully construct their contributions to online discussions about race. They know that their comments will be read by lots of people, some of whom will feel freer to critique their thoughts than they might in person.

Might it take this increased effort, soundness of argument, and relevant evidence to counter commonly held beliefs about race? The normalization of formal language, evidence, and peer critique helps establish online counterspaces as uniquely capable of facilitating resistance to dominant norms around racial discussions, including silence and colorblindness. Individual-level and ahistorical explanations for racial inequality, for example, may be less convincing in contexts where respondents can quickly share information about the structural or historical antecedents to various forms of inequality.

This shift in norms affects the way folks experience and interpret macroaggressions. Not knowing whether your experience is a microaggression, talking yourself out of thinking the experience is a microaggression, not knowing how to respond, and not being able to respond quickly enough are four of the six reasons Sue gives for people not responding to microaggressions. Online, the dynamics behind these four reasons shift. With more time to process your experience and think about a potential response, and more access to support—or people who can help you process your experience—it may be easier to both be confident in your interpretation of experiences with microaggressions, and think deeply about how you want to respond, on your own time.

Several students told me about an online incident where a White student took a picture of a university dining hall employee, a Black man, sitting down and reading the school newspaper, and posted it on a campus Facebook page captioned with "haha." Miguel, a Black student in Chicago, told me that he wrote fifteen to twenty comments on the post, believing that the original post was laughing at a Black, working-class man reading, as if this were an oxymoron. But some online users did not accept this analysis.

> People were like, "Oh no no no, you just, you just don't get it. It's cause someone is actually reading [the school newspaper], guys, like, you know, 'cause no one reads [the school newspaper]" . . . one guy was, like . . . "Oh, I definitely think that this dining hall employee can read . . . maybe the fact that [you] took such offense to this only shows that you have insecurities about him not being able to read" . . . and I was, like, Or maybe this is just racist."

It is easy to imagine how this joke might be more defensible in person. If the offending party were criticized for making a joke based on the stereotype that Black men can't read, they could easily respond by saying, *No, the joke was about the newspaper. I didn't even notice the guy reading it being Black.* A person of Color, whether or not they believed the explanation, might feel social pressure to concede that it was a misunderstanding.

Online, these types of colorblind arguments are less effective. For one thing, folks of Color feel more supported and less alone and outnumbered online, making them less likely to be intimidated into accepting that the incident was a misunderstanding. Because online racial discussions do not require participants to occupy the same physical space, nor be engaged at the same time, there is a larger pool of potential respondents to microaggressions online than in person. While it is common for students of Color at predominantly White institutions to be the only member of their racial or ethnic group in a classroom, they are never the only person of Color on the internet.

In person, microaggressions are only heard by people in the immediate vicinity, limiting the pool of people who can potentially respond and putting pressure on students of Color to be race representatives. Not all students are equally equipped or inclined to respond. Amanda says, "I know how White people are on this campus, and I know that they say things that are just blatantly racist and I wonder what happens when these like 'Uncle Tom'[15] people are around and they are just, like, 'I'm cool with that.'"

Amanda is critical of students of Color who do not counter microaggressions, but the decision to respond to racism or microaggressions is complex. In fact, at a different point in our interview (something I discuss in chapter 2), Amanda shared an example of a time that she heard a student use the n-word, but chose not to respond because she wanted to "pick her battles" and not seem like an angry Black woman. Still, her sentiment is indicative of the how students of Color think about responding to racial slights on a majority-white campus. As the only or one of a few students of Color in a classroom, students have to decide if and when to speak up against racial microaggressions made by students or faculty.

Many students shared that when race came up in class, as the only member of that race/ethnicity, the entire class would look at them as if it were their responsibility to respond. Students of Color are often saddled with pressure to choose between defending one's racial group and offending one's peers. Online, this pressure is reduced. If one person does not have the time or energy to respond to a microaggression online, someone else probably does. As Cristina says, when talking about those who make racially problematic comments online, "There's like an entire legion of people who exist to argue against these people."

In another example of collective responses to microaggressions, Jelani spoke about a White student who made online comments disparaging the neighborhood surrounding the Chicago campus and voicing how he felt unsafe with Black people walking around campus or begging for money. The online response to these types of situations, according to Jelani, is consistent: "There's never been a situation where no one has responded back to a comment like that. Someone always will step up, and it's not just like a Black person to a White person, it's like any race will comment and start saying, 'what you're expressing is, it's stupid. And, like, it's offensive. And like you really need to check yourself.'"

Let's reflect on how deep of a change in dynamic this represents. Both Cristina and Jelani *expect* people to show up when racist or microaggressive comments are made online. This is a new normal. The individual-level barriers to responding to microaggressions are lessened in online spaces, in part because of how online communication works (i.e., you can take your time in responding), but also because of the more public nature of online discussions enables collective responses to these events. It is easier to understand that a microaggression has occurred when *many* people name that action as being problematic. There is less responsibility on an individual to respond when communal responses to microaggressions are commonplace.

In the example with the picture of a Black dining hall employee reading the newspaper, after about a hundred comments, the poster said he had made a mistake and deleted his original post. While deleting the post does not necessarily signify a change of heart, it is indicative of a shift in power. For students of Color, this event is not remembered as just another time that a White student got away with a hurtful microaggression or racist joke. Instead, students collectively challenged this microaggression and denied the legitimacy of a color-blind interpretation of the joke.

Angela, a Black student in Chicago, reflected on the same incident, saying, "How did he learn that this would be an okay thing to do? . . . It reflects, in some ways, how he's been taught to think . . . they're not being challenged to not think like that." Elsewhere in her interview, Angela spoke about how seriously she takes posts on social media, believing they are indicative of how people think when their guard is down. Here, as Angela asks where the poster learned this type of behavior was okay, the answer is: anywhere. Anywhere the community does not step up to challenge that way of thinking. Anywhere folks of Color feel pressure to accept that it was an innocent statement, not a microaggression. But in some online spaces, including online counterspaces, claiming innocence does not absolve you of racial sins. Because the expression of racial stereotypes and microaggressions can go unchallenged in many on campus spaces, this incident may represent the first time the poster's racial ideology had been critiqued.

WHAT DOES IT MATTER ANYWAY?

What's the big deal? Why do these experiences matter if they're so minor? Microaggressions are *micro,* after all. The reality is, while experiences with racism may seem minor from an outsider's perspective, they might not feel minor for the people involved.

For me, the experience of being carded while heading into my own school really bothered me. It made me feel like I didn't fully belong at the school. And while I felt I got the upper hand, by *not* showing my ID and subtly hinting at how the request was problematic, I was worried about the potential ramifications. I didn't tell anyone about this incident for a long time. When I finally talked about it, it was only with close friends who I knew would keep it in confidence. I thought that saying something could hurt my status as a student or even my career down the line.

This type of worry has an emotional, psychological, and physical cost. In a survey I conducted asking questions about online and offline experiences with racism, I found that experiences with microaggressions, both in person and online, are associated with increased symptoms of depression and anxiety and increased stress.[16] These findings are consistent with the large body of research documenting the negative effects of racism or discrimination on wellness, health, and mental health, including both structural racism, such as segregation, and perceived interpersonal racism.[17] The word *perceived* here is key: it doesn't matter if others doubt that the experience in question is related to race. If a person of Color believes the experience was about racism, this is enough to predict more negative wellness outcomes including increased depression, anxiety, raised blood pressure, lower self-esteem, and even binge drinking.[18]

This doesn't mean that the problem is with people imagining racism lives where it doesn't. It means that experiences like being carded are harmful, even if the person behind the microaggression never admits racism. Marginalized people often have a window into the way oppression operates that people from privileged groups do not. By listening to the narratives of oppressed groups, therefore, we are better able to understand what subtle forms of racism and oppression, such as microaggressions, look like.

Given the high physical, social, and psychological costs of racism, for many folks of Color, not responding to racism or microaggressions may be a form of self-protection. For example, April says, "I definitely pick and choose my battles, because, like, if I had to like deal with every injustice, or every biased thing that I come across every day, I'd be utterly exhausted." Indeed, Black feminist writer and activist bell hooks discusses the choice to not to respond to racism as an act of protecting and caring for oneself.[19]

There are also consequences beyond individual well-being. In my case, there was an academic cost. This professor might otherwise have been someone I sought out as a mentor, but after this experience I was determined to avoid him. He taught a course that I was required to take in order to graduate, but I didn't take it. I was scared to sit in a classroom where he had any control over my academic success, in case he remembered the incident and held a grudge, or in case whatever preconceived notions he had about me would carry over to the way he evaluated my work or class participation. I was able to graduate without his class, fortunately, because a different professor started teaching it a few years later. If that hadn't have happened, I'm not sure what I would have done. I know I would not have taken it with him, whatever the consequence.

Perhaps that was an irrational response on my part—not the most meaningful place to make a last stand. But I wouldn't be the first student of Color to change their behavior—or leave school—in response to experiences with racism. As an undergraduate student, I conducted my first research study, hoping to understand why Black students at my school had such a low retention rate. I spoke with a sample of former Black students who had left the college over the previous fifty years (and a comparison sample of Black students who had graduated) to understand what drove the disproportionate rate of Black students leaving. Financial or familial concerns were behind some decisions to leave. But more often, folks described the racial environment at the school, or negative experiences with racism, as playing a large role in their decisions.

Experimental research on what scholars call stereotype threat shows that when people have a negative stereotype about their "group" (i.e., race, gender, orientation, etc.) reinforced directly before performing a task related to the stereotype, they are more likely to perform worse than a control group that did not have the stereotype reinforced.[20] For example, one study gave two groups of women a math exam.[21] One group was told there would be gender differences in results of the exam, while the other group was told there would be no gender differences. The group that had been told of the gender differences, thereby "activating" the stereotype that women aren't as good at math as men, performed worse than the other group. Studies have repeatedly shown that stereotype activation (alerting people to the fact that stereotypes about their group are relevant to the task at hand) predicts lower performance. When subtle racism causes people to believe that others have certain assumptions about their skills or abilities, they may use up cognitive energy worrying about the stereotype instead of their performance, and therefore be more likely to fail.

Subtle racism has other consequences for hiring, policing, and within the criminal justice system. A field experiment of low-wage employment in New York City found that not only were White applicants twice as likely to receive callbacks or job offers as equally qualified Black applicants, but also that Black and Latinx applicants were no more likely to receive callbacks than White applicants with criminal records.[22] And in a study of doctor responses to Black and White patients, who were actors presenting identical symptoms of chest pain, Whites were significantly more likely to be given follow-up tests than Blacks.[23] In the criminal justice system, a large body of research has identified the unspoken bias that results in Black and Latinx folks being stopped more often by police, given longer sentences for the same crime, and at increased risk of violence from the police.[24]

The term *institutional racism* comes from Black Power organizer Kwame Ture (formerly Stokely Carmichael) and Charles Hamilton who use it to describe covert systems that privilege the White community over the Black community and are harder to detect than individual-level racism.[25] The above examples of structural, systemic, or institutional racism point to systems that maintain racial inequality covertly, without ever needing to be openly racist. Knowing that subtle experiences with racism, such as microaggressions, are harmful for folks of

Color, should be enough of a reason to take them seriously. But understanding that these experiences are the interpersonal manifestations of institutional racism—the ways that racial boundaries, and racial inequality are reproduced and maintained—should increase our sense of urgency and make us take subtle racism as seriously as overt racism.

Rachel gives an example of a subtle microaggression, also posted on a campus website, that is linked to systemic racism and the brutal treatment of Black folks in the criminal justice system.

> This one girl posted, like, "Overthought while walking behind a black guy, what if I robbed him to just to be ironic?" And someone was like, "What? . . . how is that ironic? Like, so you're admitting that you expect a black guy would rob you?"

Like the post of the dining hall worker, this post garnered a large number of comments. In this instance, however, the post was deleted by the student moderator of the campus page for being offensive—something that happens less often for racial slights but is more likely to happen when a post is overtly offensive. Rachel noted that the consequences of the post extended to offline life, and said, "For the next couple of weeks when everybody saw that girl walking around on the quad, they're like, 'It's her!'"

Implying the criminality of persons of Color is a textbook microaggression. There is also an obvious link between this interpersonal microaggression and structural racism in the form of mass incarceration and anti-Black police and vigilante violence. If you are always scared Black people around you are there to rob or rape you, you're more likely to call the police on them or worse. The assumption made into a joke by the user who wrote this post, is the exact same assumption Trayvon Martin's killer made when he chased down the unarmed Black teenager and murdered him in cold blood.

Here Rachel talks about the way students of Color created their own system of peer-based accountability that began online and then carried over to campus. In an academic paper I wrote about this, I called this phenomenon *online racial checking,* which refers to "critical responses to microaggressions or the expression of racist ideologies."[26] Accountability is a part of online racial checking, but otherwise is rare on college campuses, where administrators may make statements about overtly Racist incidents that have the potential to be newsworthy and embarrassing, but are more likely to ignore minor issues such as microaggressions.

In this example, as the poster was being ostracized by many on campus, an activist of Color approached her in person to talk about the incident. The poster acknowledged how she was wrong and showed interest in increasing her understanding of race and racism. Later that academic year she came with the activist to a Black student party, where her interactions with students of Color indicated to Rachel that she had changed the way she thought about race.

Believing that responding to a microaggression is useless and won't change anything is one of Sue's six reasons that people don't respond to microaggressions. This example shows that online racial checking can sometimes lead to transformation, admission of wrongdoing, and genuine change. And as people witness the effectiveness of online racial checking, it may become more common.

Sometimes online racial checking is less confrontational, though no less important. As Natasha, a biracial student (Southeast Asian and White) in Los Angeles points out, online responses to microaggressions do not always assume the poster is racist: "It could be, like, something they didn't know was offensive, and someone will be like, 'Hey, that's offensive,' and they'll be, 'Oh okay, let me change my statement,' or 'Let me delete my statement.'"

Any of us, even those who identify as antiracist, can make mistakes that cause harm. When I teach classes that deal with racism, some students are scared to talk because they don't want to say the wrong thing and "sound racist." And sometimes students of Color may be worried about being harmed by what well-intentioned White folks say. It is a challenge to create an environment where people are willing to engage in uncomfortable conversations, where they are at risk of both causing harm and being harmed. One of the ways I try to do this is by setting the expectation that any of us, myself included, can and should be challenged if we say something hurtful, even unintentionally. Establishing classroom expectations that normalize critique of ideas, and normalize humility in responding to challenge, is difficult. Some of the norms in online communication, including being able to step away and gather oneself, and not feeling the pressure of being attacked in the moment in a face-to-face setting, make this easier.

Online racial checking may not work every time. But witnessing the effectiveness of online racial checking may make folks of Color more optimistic about whether challenging racism and microaggression online can make a difference in the real world—both online and offline. This may seem a small intervention. But remember that, most of the time, this stereotype goes unchallenged. It is taken for granted, and it is met with silence or laughter. Every time microaggressions like this go unchecked, they may be internalized by Black folks in a way that harms their health, and internalized by White folks in a way that may make them more likely to call the police on innocent people. Think about the "Karen" epidemic, as videos of White people calling the police on Black people for no reason have gone viral over the past few years. In Rachel's example, this interpersonal intervention stopped the reproduction of a racist ideology that is often taken for granted— that Black people are criminal. Challenges like this can weaken the justification of racist policing practices—or even faculty deciding to card students of Color on campus. Responding like this is resistance.

HOW SHOULD I RESPOND?

When people of Color share these types of experiences with each other, we often ask each other, "Well, what did you do?" Did you shut them down? Curse them out? Physically hit them? Report them to HR?" When the answer to any of those questions is yes, the person gets some "woke" points. You did it! You didn't take racism without doing anything about it. You stood up for yourself. You stood up for us.

More often, however, the answer is no—I didn't do any of those things. Because I want to keep my job. Because, while I'm hurt by what they said, I know they mean well and don't want to lose that friendship. Because I didn't want them to spit in my food. Because they walked away too fast. Because takeoff was in fifteen minutes, and I didn't want to risk being asked to leave the plane. Because I didn't want a bad grade.

Even when folks of Color decide to respond, it isn't easy. Think about the example of my friend JT getting carded at the law school for a few weeks. He eventually confronted the security guard, an act not everyone would be comfortable with. But, even in this case, he only confronted them after weeks of being stopped—that's weeks of reflecting on the experience, processing the hurt, confusion, and maybe anger—and finally making the decision to say something. For many people who experience masked racism, these experiences are fleeting. You have only a split second to respond, and if you don't do it right away, you may never get another chance. Typically, folks of Color don't experience microaggressions from the same person in the same way over and over, at predictable times. This means they might have fewer opportunities to stand up for themselves.

People tend to *overestimate* the extent to which they would respond in a racially problematic scenario. To measure this, researchers conducted an experiment, where one group of participants was asked whether they would respond to a hypothetical racist act if it were to happen in front of them. Another group was shown the racist act, believing it to be real, and unaware that their reactions were being studied.[27] Many more people said they would react to the hypothetical racism than actually reacted when given the chance. Most of the time, it seems, when a microaggression or subtle act of racism takes place, no one speaks up about it. People of Color may decide to share these experiences when they feel safe, but rarely challenge those events as they occur in practice. People of Color may want to avoid a reputation for being too sensitive to issues of race, or could worry about losing friendships by calling out their peers for racially problematic statements or behaviors.

Jason, a Black student in Boston, told me a story about being micro-aggressed upon by a professor during class. Some White students in the class stood up for him, raising their hands to tell the professor how they were engaging in problematic behavior. Jason felt validated by their support, but regretted it later when the professor singled him out for out-of-class disciplinary action. The stakes are high, and people of Color know that challenging racism or microaggressions can negatively impact their lives. This is another of Sue's reason's people don't respond: knowing the potential negative consequences of their actions.

Given the potential ramifications, not responding to microaggressions is actually a very reasonable response. People who study and engage in racial discourse often find that White folks frequently don't respond well to challenges about race or racism. As we discuss earlier in the book, White fragility is built around White defensiveness and hypersensitivity to any suggestions that the White individual may be implicated in racism.[28] If your classmates, co-workers, friends, or family will take offense at the idea that they hurt you with a microaggression, you may be less likely to bring it up. Especially given how ostensibly ambiguous or easily defensible subtle acts of racism are. *No, I didn't mean it that way; I wasn't thinking about race!* Without access to someone's inner dialogue, how is one to suggest otherwise?

Jeremiah, a Black and Latinx student in Chicago, relayed a story that began when he used the word "doe," slang for "though," in conversation on campus. A White student responded to him, saying, "'I didn't know we were talking about deer, here,' you know, and then he was just like, 'Well, I wasn't talking about deer. I was very clearly using a form of slang, and that's how language works.'"

When Jeremiah attempted to engage this student in conversation about the incident, the student simply walked away. At this point in the interview, Jeremiah took out his laptop and read me the online post he wrote about the incident, which had garnered over a hundred comments in a week's time. I told him that the post read like an essay, and he responded:

> I like to write in a very provocative manner, so I write it—I sit down, I have a topic, or I have an article, and I'll put my opinions about it and then I get to writing about it in a way I know people are going to come talk to me about it for. Certain people that either agree with me are going to say, "I agree, but this is what I don't agree with," or certain people are going to say, "I don't agree with you at all," and that's where the conversation happens. These specific posts usually are written in a very long prose manner . . . the ones that garner the most attention are very—are written in, like, an intricate manner.

One limitation of responding to racism or microaggressions in person is that people may be unwilling to listen to what you have to say. But by turning to the internet to write a detailed, articulate, and critical post about the incident, Jeremiah increased his agency. He started a public discussion that went beyond his personal experience with a microaggression, and that extended to a conversation about the relationship between race and speech. The responses to his post were mixed, with some support and some critique of his interpretation. But whether or not people who commented on his post agreed with him, the post prevented the microaggression from being further ingrained in campus life as normal and acceptable, something to be ignored and walked away from.

This is the type of incident that other students may share in traditional counterspaces, where they receive moral support from other students of Color. But by engaging in online racial checking, Jeremiah created an online counterspace where he both had immediate access to personal support and increased his personal agency by engaging both White students and students of Color in a discussion that critiqued the prevalence of racial microaggressions on campus. This is an action that has the potential to shift the racial climate on campus.

RESISTANCE

Most research on microaggressions is devoted to understanding the ways they impact people of Color and how people of Color cope with these experiences. Although research on resistance to microaggressions is limited, researchers have explored maintaining silence in the face of microaggressions as a self-protective practice; private resistance behind closed doors or in safe spaces (such as counterspaces); and organized resistance through protest or activism in response to prolonged or acute experiences with microaggressions.[29] Highlighting these multiple forms of resistance shows that targets of microaggressions are not victims without agency, persons who are being acted on but who themselves do not engage in meaningful action.

In this chapter, we have been exploring forms of *active* resistance against microaggressions, a phenomenon that has not been studied much in the scholarly literature. Technology is central to this story. Students of Color think differently about responding to racism online versus in person, and the social norms that make responding to racism awkward in face-to-face situations are lessened or even reversed online. Internet-based communication, such as social media, gives students access to unique tools that increase their perceived and actual capacity to respond critically to racial microaggressions. As technology changes the way regular people respond to regular, everyday events, the result is anything but regular. It's radical and empowering, and shifts racial power dynamics.

The decreased power differential between Whites and students of Color in online counterspaces can lead to Whites facing increased peer-accountability for making racial slights. Rachel speaks about the potential consequences of making offensive posts online.

> If someone says something really racist, you get a message like, "Did you see what she said? She's in my such-and-such class." And then, they get "side-eyed" until you actually bring it up to them . . . That's a very, very easy way to be ostracized . . . you're not going to be able to still be accepted if you post something racist or homophobic.

This outcome, Whites being made to feel uncomfortable on campus, or being less accepted because of their views or comments on race, is different from what we have come to expect on college campuses. Students of Color in online counterspaces have access to informal sanctioning mechanisms that not only punish perpetrators of racial slights but may also discourage further racist or homophobic comments.

In another example, Stephanie, a Black student in Chicago, talks about a student who posted an article about Harriet Tubman being put on the twenty-dollar bill. A White student responded that he was going to miss Andrew Jackson. A flurry of comments ensued, and many students pointed out how, among other morally questionable actions, Andrew Jackson had been involved in a genocide, and therefore was not a historical figure who should be honored by having his face on our currency. The student continued to defend Andrew Jackson, and Stephanie says of her own response, "I commented something, like, oh my god, I can't even remember what this comment was, but it was—it was gold, it was really good . . . a whole bunch of people liked it and I felt validated by the fact that a whole bunch of people liked my, like, shaming of this guy."

The reality that a White man can be shamed because he defends a former United States president, and that this shaming garners much support (in the form of Facebook likes), is a shift in how we have traditionally thought about racial dynamics on college campuses. By engaging in online racial checking, Stephanie participated in the creation of an online counterspace where dominant historical narratives were questioned. She does not remember the words she used in her post, but she remembers the social impact it had, the support she received, and how in that moment the campus racial power dynamics were turned upside down.

DIGITAL SPACES AND NEW OPPORTUNITIES TO RESIST

The quote at the beginning of this chapter comes from a book in one of my favorite fantasy series, *The Wheel of Time,* as a teacher describes the swagger that seems to be associated with people who learn they have the ability to wield magic. Traditional, respectable power dynamics matter less to these students as they realize they are capable of changing the world. Can resistance be like magic in this way? It empowers folks of Color, reminds us that oppression is not permanent, and demonstrates our ability to change the norms, behaviors, and policies that limit our lives. As we learn to resist, expect us to walk differently—head and shoulders high, unburdened by the perceived need to live according to the standards and limitations established by the oppressor—or the rules of racial engagement.

There is also emerging evidence that a dose of resistance may be good for your health. Earlier in the chapter I refer to the survey I collected that showed that experiences with racism were associated with increased symptoms of depression, anxiety, and stress. But with the same data, I found that resistance can reduce the harmful effects of racism. For people who report witnessing others respond to racism or responding themselves more frequently, or who report posting online about their in-person experiences with racism, the negative effects of experiences with microaggressions on depression, anxiety, and stress are lessened.[30]

As colleges and universities around the country decide how to navigate the thin line between free speech and harassment, students of

Color may not rely on official policies to determine the way that they respond to racial microaggressions. Online counterspaces and online racial checking increase the agency students of Color have on college campuses in determining the way race is understood and experienced. Students of Color report that White students who commit racial microaggressions in person, intentionally or unintentionally, are rarely challenged. But in online counterspaces, students of Color are more comfortable engaging in online racial checking than they are responding to racial microaggressions on campus.

Students of Color are not indefensible. With the tools available in online spaces, students of Color are shifting the meaning of race, and are themselves deciding what types of racial jokes and comments are acceptable or unacceptable. In the absence of an institutional push against microaggressions, or a structured plan for racial learning for all students, students of Color are leading the way in implementing informal sanctions against microaggressors, and creating online counterspaces where dominant racial ideologies are questioned.

This, of course, can make White students who engage in subtle forms of racism uncomfortable. Might we expect a backlash to these changing norms that highlight and challenge behaviors and ways of discussing race that have, for so long, been seen as being normal and innocuous?

Perhaps. On Twitter, conservative thinker Nate Hochman shared a finding from his research for a forthcoming *Atlantic* article, saying, "Many young conservatives can point to a 'radicalization moment.' There's an analogue to this on the left: young leftists often say they were radicalized by the 2008 recession/failures of Obama admin/etc. For young cons, it often revolves around experiences on campus/social media."

Can challenging racism online radicalize the right?[31] Should people of Color worry about "creating" racists by hurting the feelings of White folks who are offended by being held to account for hurtful things they say or do? Making White people uncomfortable does break one of the rules of racial discourse we discuss in chapter 3, after all.

Students should not have to preserve the feelings of people who engage in behaviors or speech that cause harm—but they tend to anyway. The intentionality with which student activists engage people whom they seek to help grow, anticipating the type of defensiveness, sensitivity, and fragility that characterizes many conversations about race, is something that was visible in the interviews in this chapter, and it is something we explore more in chapter 7. However carefully they choose to engage, however, some White people will not take kindly to being confronted.

It should not be surprising that the change in power dynamics that enables folks of Color to step outside of the rules of racial discourse, challenging racism and sometimes making White people uncomfortable, may have repercussions. Some White folks who are challenged will grow, like the young woman we discuss in this chapter, who attended the Black party with an activist. Others, like those Hochman refers to, may double down on their biases and fight for their right to be privileged (or racist) and proud—no matter how carefully students engage them in critical discourse. It's clear to me, however, that if witnessing antiracist discourse online is enough to radicalize a young conservative, the ideology was already there and just needed a reason (or excuse) to emerge.

Remember that racial ideologies adapt to systems of domination. When White folks are able to maintain superiority through everyday colorblind racism, and without being challenged, there is no need for radicalization or open racism. But when marginalized folks refuse to succumb to oppression, challenging oppressive rules, systems, and interactions, this can create discomfort for the ruling class. Some people in the dominant group will use their discomfort as a call to arms and fight harder to maintain their position of power.

The pathway to freedom has never been meekly asking the oppressor to relinquish his grip. As we demand and fight for our freedom, racists may hold on to their positions harder and more openly. This potential, however, should not dissuade us from the task of naming, condemning, and revolting against White supremacy in all its forms.

There is still hope in resistance. My colleagues and I conducted a study where we tested the effectiveness of online messaging around race on creating user change, or what we called "making concessions."[32] In 2018 Nike ran an ad featuring ex-NFL player Colin Kaepernick, who lost his job playing football due to his kneeling during the national anthem in protest of anti-Black police violence. As anti-Kaepernick users on Twitter were burning their Nike gear in frustration with this ad, we gathered a little over two million tweets about Colin Kaepernick over a one-week period and followed a sample of users, looking for those who signaled a willingness to make a concession during this period. An example of a concession would be "if a user tweeted that Kaepernick shouldn't have kneeled down during the national anthem because it was disrespectful to the flag, and then in the later conversations with other users this user acknowledged that Kaepernick has the right and a reasonable cause to protest."[33] We found that one of the variables that predicted increased likelihood of making concessions was the number of messages Twitter users received on the subject. Users who were challenged or engaged more often were more likely to signal a willingness to consider a different point of view.

Many people assume that online conversations about race or politics are pointless, especially given how they tend to become uncivil. But our study shows that there is potential for online racial checking to influence how people think about issues of race. Not everyone who witnesses or is on the receiving end of digital resistance will become an antiracist. But there is meaning in resistance, empowering and supporting folks of Color, challenging dominant norms and structures, and even creating opportunities for growth and change.

The alternative to challenging racism is accepting the racial status quo. For people of Color, racism is poison and accepting it is not an option. In the next chapter we leave the college campus to explore resistance on Twitter, as users disrupt the standard pathways of racism—or the ways it effects how people of Color view themselves—and renegotiate what it means to be a person of Color in a racialized world.

6

DOUBLE-SIDED CONSCIOUSNESS

The Negro is a sort of seventh son, born with a veil, and gifted with second-sight in this American world,—a world which yields him no true self-consciousness, but only lets him see himself through the revelation of the other world. It is a peculiar sensation, this double-consciousness, this sense of always looking at one's self through the eyes of others, of measuring one's soul by the tape of a world that looks on in amused contempt and pity.

—W. E. B. DU BOIS, *The Souls of Black Folk*, 1903

KENDRICK LAMAR IS ONE of my favorite rappers. His music is intelligent without being pretentious and streetwise without glorifying the violence that results from racist public policy and the lack of opportunity that characterizes some poor, Black communities. It's also fresh. Kendrick consistently comes up with new and unique sounds, each album with a unique theme and sonic thumbprint. His music pleases the hip hop heads who pine for the golden age, plus it sells—and those two don't always coincide. To top it off, he and I were born on the same day: June 17, 1987. I don't have any platinum records to boast of and I'm no astrologer, but I like to think something was in the stars on 6/17/87. Maybe Mars was bright that night.

I once watched a video of Kendrick Lamar performing his song "m.A.A.d city" in front of a live audience. The lyrics were:

Man down . . . Where you from, nigga?
Fuck who you know, where you from, my nigga?
Where your grandma stay, huh, my nigga?
This m.A.A.d city I run, my nigga

I don't have a problem with Black artists using the n-word. But this video astounded me. Because as Kendrick rapped, the entire crowd—mostly White people, jumping up and down to the beat—sang along. At some point, the voices of the crowd drowned out the music, as hundreds, or maybe thousands, of White people chanted the n-word, over and over. Despite being far removed from the video, I felt the hair on my arms stand up and I shivered a little. White people using the n-word, for me (and many other Black people), brings up feelings of danger.

I wondered—has the n-word ever been said by this many White people in one room, this many times, in just ten seconds? Kendrick is what you call a conscious rapper. He's unquestionably pro-Black. But here, at his concert, he may have broken the world record for the largest number of White people shouting a racial slur. What does intelligent, fist-raised, and conscious Kendrick think about this? I was just watching a short clip, but it was Kendrick's job to do this every night as he gets paid millions of dollars to perform for mostly White crowds. I imagine Kendrick would need to either rationalize this practice and be okay with Whites saying the n-word with his lyrics, or, it would eat him up inside.

For my birthday in 2017, my friend JT took me to a Kendrick concert. It was incredible, and like many others at the show, I sang along with most of the songs. But when "m.A.A.d. city" played, I stopped singing. I looked around the United Center stadium in Chicago, watching all the White people to see if I would witness what I saw in that video. I held my breath, but thankfully I didn't see one White person saying the n-word along with Kendrick—not one! I was able to exhale. I'm not sure how to explain the difference between my experience and the video I had seen.[1] Maybe we weren't in a rowdy section, or maybe that happens at university concerts and not stadium concerts. It could even be that at some point Kendrick intentionally did something to prevent this from happening at future shows.

In 2018, years after I watched the video of the White crowd chanting the n-word, Kendrick invited a White fan onstage to rap his lyrics in front of the crowd, but stopped and chastised her when she said the n-word along with his song.[2] Perhaps it was easier to respond to one person saying the n-word than a crowd of thousands, or perhaps his thinking about the n-word changed. Four years later, in 2022, Kendrick reflected on this fan incident in his song "Auntie Diaries" as he told the story of a transgender family member who challenged Kendrick's "playful" usage of an antigay slur (the f-word) and suggested Kenrick could continue to use it only if he was also okay with a White woman at his concert using the n-word. The implication was that Kendrick realized neither was okay, and the song explored his move away from cultural and religious homophobia and heteronormativity and toward radical acceptance and queer allyship.

Another of my favorite rappers, Vince Staples, has some lyrics that reflect on the racial dynamics of the n-word and his mostly White concert audiences. In the second verse of his song "Lift me up," Vince says:

All these White folks chanting when I asked 'em where my niggas at
Goin' crazy, got me goin' crazy, I can't get wit' that
Wonder if they know, I know they won't go where we kick it at
Ho, this shit ain't Gryffindor, we really killin', kickin' doors

Here Vince references the White crowds going crazy at his concerts. Whether he's referring to White people saying "nigga" along with him, or just responding to messages that were intended for a particular audience (i.e., a message to Black people, or to those involved in street life), he's uncomfortable with his audience assuming the message or language is for them. Vince also references a personal, psychic cost; the actions of his White fans in this case make him feel like he's losing his sanity.

In the next line, Vince questions whether the Whites he mentions understand his inner thoughts. He suggests that he knows what they, White folks, will or won't do. But they don't understand what's going on in his head. And then in the last line of this stanza, Vince tells the White audience, "this shit ain't Gryffindor," referencing a House in Hogwarts, the fictional setting of the *Harry Potter* book and film series. The street shit that Vince is kicking, when he asks "where my niggas at," when he says "we really killin'," is real, not fantasy like wizards and centaurs. Do White fans know that these lyrics are explorations of Vince's experiences as a member of a gang, and perceptions of the urban blight in his hometown of Long Beach, California?

They don't, presumably, and instead simply experience the music as pure fun. For them, the violence Vince references is like the violence in a film—the gore adds to the experience, it raises your heart rate, it makes the scene believable—and there's no thought about the real human beings who have experienced that type of violence and its consequences.

W. E. B. Du Bois wrote about a metaphorical "veil" that separates the world as experienced by Black people and that experienced by White people. He says that White people can only see their side of the veil—it obscures from them the realities of what it means to be Black. Black people, on the other hand, must understand the White world in order to survive, and are therefore able to see through the veil, understanding both Black and White realities. Vince's knowledge of how Whites experience his music, and his understanding of the parts of his music they are incapable of grasping, are an example of what Du Bois calls double consciousness.

Few theories of race have been as widely taught and discussed as Du Bois's double consciousness. At its core, double consciousness refers to a contradiction that Du Bois sees as being endemic to Black people in the United States: they are American, but they are denied full access to citizenship and are subject to discrimination at state, institutional, and individual levels. Black folk living within this contradiction—or behind the veil—are hyperaware of the way they are looked down upon by Whites. Du Bois suggests that this awareness precludes the ability of Blacks to maintain a singular self-consciousness, as their own self-concept is continuously challenged by their knowledge of White prejudice and disdain. The result is double consciousness, a problematic psychological state that saps emotional energy and prevents Black people from being whole in an almost spiritual sense.

While DuBois frames double consciousness as a widespread problem plaguing the Black community, he sees it as a private, internal struggle, something that is universally experienced but rarely articulated. He therefore proposes an individual-level solution to the double-consciousness problem—a melding of the double consciousness to form a single consciousness. It is unclear, however, what this coalescing process might entail or how the product of a fused consciousness might appear or function in a racialized society. Perhaps Du Bois believed that double consciousness could only be resolved *after* the nation solved what he famously called "the problem of the color line."[3]

While Du Bois's original conception of double consciousness referenced a struggle unique to African Americans in the United States, scholars have applied his theory to oppressed groups in other contexts.[4] We can therefore understand the color line that Du Bois foresaw as the biggest problem of the twentieth century, as not only a Black and White color line but also the line between colonizing countries[5] and the colonized peoples of Africa, Asia, Australia, the Caribbean, and the Americas.[6]

The problem of the color line—racism and racial hierarchy—has certainly not been resolved. Neither has the problem of double consciousness. Many scholars, writers, and educators have used Du Bois's theory to explain how Black people, and other groups of Color, experience racism and life in a racialized world. Other research does not explicitly invoke Du Bois's theory, but explores conceptually similar internal experiences with and responses to White racial attitudes. For example, as we discuss in the last chapter, research on stereotype threat finds that when people of Color are exposed to stereotypes, they perform worse; knowing that people think less of you can reduce your capability on a variety of tasks.[7] Similarly, a large body of research on racial identity finds that how you believe others think about your race can shape how you cope with racism.[8] And, of course, research on microaggressions explores the effects of subtle racial slights that fall short of explicit racism but still communicate and reinforce marginalized status to people of Color.[9] These bodies of work elucidate some of the ways double consciousness can be reinforced and reproduced in interactions that communicate inferiority—in other words, the ways awareness of White prejudice can have an adverse effect on people of Color.

I want to come at the double-consciousness problem from a different direction. What if we shift the focus from individual-level responses to public and communal engagement with racism? What happens when, instead of struggling alone to live with the weight of knowing what White people think about them, Black folk and other people of Color collectively expose, critique, and reject the manifestations of White supremacy, both structural and interpersonal, that give rise to and reproduce double consciousness?

As we see in the last chapter, technology enables this shift. As racism is unmasked, as resistance is digitized, folks of Color are inverting the veil—shining the light on Whiteness and exposing the destructive spirit of White supremacy—and in so doing are creating a new pathway for confronting the durable double-consciousness problem.

The result is what I call *double-sided consciousness*. Instead of people of Color looking at themselves through the eyes of others (i.e., White people), double-sided consciousness refers to a reversing of the double-consciousness mindset through a public and communal project that reveals what White people look like—from cultural practices to the manifestations of privilege—through the eyes of people of Color. Double consciousness is a state experienced by Black people as a result of their experiences and knowledge of racism, a result of the pathologizing of Blackness. Double-sided consciousness, on the other hand, is projected by Black people and people of Color as they switch the focus to Whiteness, as well as problematize White supremacy.

As an example, take a look at the following messages (or tweets) posted on Twitter:

> Wypipo deeply care about being called a racist, but not about their actions that make them one.

> Wypipo genuinely care about animals more than people & it actually blows me

> i'm here for the wypipo who dont step in and say "not all wypipo" when im talking about things wypipo do. thank u

These messages use the term, "Wypipo," a creative spelling that, when read aloud sounds like the words "White people." Much more than shorthand for White people, however, the term *Wypipo* is used as a way of communicating experiences with, feelings about, and knowledge of the ways Whites think and act in a racialized world. In the above examples, Twitter users are making generalizations about White behaviors, sentiments, concerns, or spaces.

These tweets represent an othering of Whiteness. Whereas Whiteness is often thought to be invisible, and to refer to normal, standard ways of being, here Whiteness is named and defined. Sometimes this is derogatory, sometimes humorous, with the humor often being based around perceived collective experiences with White people. In some instances, tweets about Wypipo represent a reversal of stereotypes.

> wypipo talk loud af☺I'm aggravated

> I got scared to party with wypipo when I first went to one of their parties in college and they had lines of coke in the table smh

In chapter 4 and 5, the opposite messages were posted to the PIC and a campus student page, suggesting that Black students were too loud for the library, or were criminal and to be feared. These Wypipo tweets quite literally flip the script.

To demonstrate and explore double-sided consciousness, I leave the college campus and head to Twitter. The way the term *Wypipo* is used exemplifies the dynamic changes in racial discourse that occur as racism is unmasked in online spaces, as resistance is digitized, and marginalized people have access to new antiracist tools to invert the veil, shedding the light on Whiteness, and potentially transforming the way Black people experience and respond to the double-consciousness problem.

I used a data-scraping program to collect every tweet using the term *Wypipo* between 2015 and 2021—over 1.1 million tweets—to see how the usage changed over time. I also took a random sample of around a thousand of the most influential tweets in order to take an in-depth look at these for analysis. While the term *Wypipo* was barely used at the start of 2015, by the end of the year there were more than one thousand original messages using the term each month, demonstrating how quickly the neologism became widely used on Twitter. In figure A we can see usage of the term rising steadily until it leveled off in 2018. The spike in total mentions around August or September 2017 seems to coincide with the Unite the Right White supremacist rally in Charlottesville, Virginia; Trump's statement that there was "blame on both sides," and "very fine people on both sides" in the wake of the rally, at which a White woman and counterprotester, Heather Heyer, was murdered by a White supremacist; and the uproar after journalist Jemele Hill called Trump a White supremacist on Twitter. These are all events that can be associated with increased racialized discussions on social media.[10]

Wypipo is a term that originated on Black Twitter. For Tressie McMillan Cottom, Black Twitter is "not a place or a group of people but a set of communication practices . . . a group of knowledges . . . shared

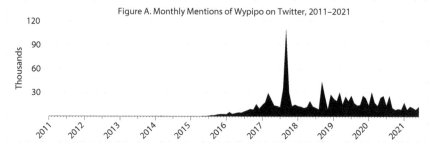

Figure A. Monthly Mentions of Wypipo on Twitter, 2011–2021

Figure A. Monthly Mentions of Wypipo on Twitter, 2011–2021

language, culture, and references."[11] While Black Twitter has become famous for its innovative strategies for resistance, from the use of hashtags such as #BlackLivesMatter and #SayHerName and social media activism, to callout culture, which Brock defines as "gendered Black cultural critique," not everyone on Black Twitter is there to resist racism.[12] Brock's book, in fact, focuses on how Black folks use the platform to celebrate and engage in topics around everyday life, from talking about work, love, and music to live tweeting during TV shows. The latter is what forced me to start watching *Game of Thrones* on time every week, because if I started it late the plot would be ruined with spoilers popping up in my Twitter notifications (#thronesyall).

This framing of Black Twitter is consistent with how Cathy Cohen thinks about resistance and the politics of deviance.[13] Oppressed peoples can engage in actions that do resistance work—challenging dominant norms, narratives, and refusing to act as the oppressed are meant to act—without these actions being intended as resistance. Instead, they may be acts of survival or just efforts at living life to the fullest, free from bonds.

That's what's going on here. For example, the term *Wypipo* ostensibly began as a way of talking about "White people" without everyone knowing, or without moderators or key word searches revealing these conversations. These tweets tell a bit about the etymology of the term:

Wypipo has been flagged, we need something new

Seeing NBPOC using "wypipo" makes me feel weird but I saw a white person using it and that's a flat NOPE.

This term is a reference to a broad critique of Whiteness. Its usage is consistent with Du Bois's suggestion that Black people can see and understand the White world, but White people cannot see through the veil to understand the realities of what it means to live as a Black person in a colonized world. Given that use of the term *Wypipo* implies a kind of second sight—the ability to understand both sides of the veil (i.e., both Whiteness and Blackness)—the user behind the second tweet here believes it would be problematic for White people, who according to Du Bois do not have second sight, to use the term *Wypipo*. Things that trend on Black Twitter tend to trend everywhere, driving the way we think about race and representation.[14] Terms endemic to Black Twitter, such as *Wypipo,* therefore, are often used by White folks or other groups of Color (i.e., Non-Black People of Color, or *NBPOC*). Rashawn Ray and colleagues have written about how using the Black Lives Matter hashtag is a form of collective identity. *Wypipo* is not a hashtag, but using the neologism may have a similar effect. *Wypipo* is a collective Black discursive project that engenders double-sided consciousness for Black folks and other groups of Color.

More than simply sharing thoughts and experiences from beyond the veil, double-sided consciousness alters the dynamics of the double-consciousness problem. As users engage in the communal rejection of the norms, behaviors, and structures that hold up White supremacy and produce double consciousness in Blacks and people of Color, this collective venting relieves the pressure of double consciousness for people of Color and puts pressure on the mechanisms of White supremacy that are often invisible or ignored. For example, see the following tweets around Twitter user experiences with Whiteness or racism:

Dear White People (wypipo): When I am ordering drinks at the bar from you (or doing anything else, really), please ffs do not end every last sentence to me with "brotha". It's not cool—it's patronizing AF.

Wypipo: MLK Said we shouldnt see color
Us: well no but he DID say white moderates are the perpetuators of
   racis . . .
Wypipo: HE HAD A DREAM!

These tweets name or call out White behaviors that the users behind them find to be problematic. Calling out microaggressive norms at bars is a form of resistance that, as discussed at length in chapter 5, may be more welcome (and safe) online than in the bar, and can expose and slow the reproduction of everyday forms of racism. The second tweet challenges dominant narratives about civil rights leaders, which can include the gentling of Martin Luther King's message, ignoring his radical beliefs, vilifying Malcolm X as if he were a terrorist and not a brilliant thinker and fearless defender, and erasing women and queer leaders like Ella Baker and Bayard Rustin who shaped the movement. In an article critiquing and challenging the dominant narrative of the civil rights movement, Jacquelyn Dowd Hall writes that "the movement's meaning has been distorted and reified by a New Right bent on reversing its gain." We can therefore understand the dominant narrative of the civil rights movement to be an intentional misrepresentation of history.

On Martin Luther King Day in 2021, my Twitter feed was filled with posts that showcased MLK's radical beliefs and challenged the version of MLK that Republican senators like to tweet about—the standard version that is taught in schools, that frames him as a hero who fought for a form of racial equality that centers individual rights and freedoms—the version that Arizona leaders used to justify ending an ethnic studies program.[15] These tweets, like the Wypipo tweets listed above, rejected White, dominant messaging about Black history and the fight for freedom. The schoolbooks tend to tell a top-down story of the civil rights movement, saying the battle against racism was waged by men and elites, and ended in 1964 with the passage of the Civil Rights Act. If that were the case, why are we not yet free? Is it due to our personal failings? Or is the willful manipulation of history that emphasizes individual choice and freedoms meant to turn our attention away from the durability of systemic racism and injustice?

Double consciousness emerges from a hyperawareness of racism and internalization of the permanence of our subdued, marginalized status. Double-sided consciousness externalizes and rejects—publicly—the mechanisms of oppression. Because double consciousness is a manifestation of experiences with White supremacy (a systemic and societal-level problem), individual-level, internalized solutions have proven difficult. Double-sided consciousness may improve our ability to combat double consciousness by moving the struggle from an individual's mind to a collective and public struggle against dominant norms, myths, narratives, and racist behaviors.

One form this can take is the discussions of actions, real or hypothetical, taken against White people. As seen in the following tweets, this discussion seems to be for some users a mixture of the cathartic and empowering, as these actions were often framed as being in responses to perceived injustices, racism, or microaggressions.

> I really don't care about making wypipo uncomfortable because majority of them are completely unaware of their privilege.

> Anyway idk ever since I started working here I've experienced so much racism and it's really testing me bc these wypipo don't want this.

This willingness to make Whites uncomfortable is noteworthy, because so much of the literature focuses on how people of Color go out of their way to make Whites feel comfortable. In chapter 3, we discuss this as being one of the rules of racial discourse in many face-to-face settings. Even after experiencing racism, many people of Color stay quiet in order to not offend Whites with accusations of racism.[16] Intentionally creating discomfort for White folks, then, may play a role in revealing various manifestations of White privilege.

Central to the double-consciousness problem is the notion that the conflict is internalized. During Du Bois's time (and still today, in some contexts), externalizing responses to racism, especially aggressive responses, can be dangerous for Black people. As we discuss in chapter 3, there are rules for how Black people and people of Color can engage with White people. People of Color perceive, understand, and typically abide by these rules. Those that don't may suffer socially, academically, and in their workplaces, and may even have negative encounters with the police if they act in ways that are deemed threatening. These discussions of actions taken against White people, then, are public declarations of the types of sentiment that are typically only "safe" behind closed doors, in counterspaces, or on the Black side of the veil. Double-sided consciousness entails the movement of these sentiments to the mainstream, as social media posts originating on Black Twitter are immediately visible on both sides of the veil.

Double-sided consciousness can also represent the promise of the "second-sight"—the ability to see the hidden mechanisms of oppression and domination—that Du Bois suggests Black people are gifted with. Only, instead of revealing this second sight in private, safe spaces, people of Color engage in these discussions in the open through public online discourse. This can include placing contemporary events in historical context or explaining why certain speech or behavior is problematic or offensive. Some tweets that do this read as intentional efforts at building consciousness among followers.

Wypipo will do all they can to promote the lie that a bunch of toothless Appalachians are the racists and klan in this country.

Too often, racism and white supremacy is discussed by wypipo as an abstract, with distance, a real problem, but one they're exempt from.

[@user][17] Ain't no levels to this. You wanna focus on individuals so you can ignore the systemic shit wypipo benefit from whether they want it or not

The first two tweets here make one of the points I've been trying to hammer home in this book: that racism is not limited to capital R, overtly Racist acts that most people seek to distance themselves from, but can also include systemic biases that are difficult to identify, or the subtle biases implicit in friendly interactions. Remember that the true definition of racism includes both an attitudinal and structural component; nonhostile attitudes that legitimate the oppressive racial order are racist. Rather than simply accept the dominant, simplistic definitions of racism, or the ways White folks can distance themselves from racism using White intellectual alibis, here users bring their definitions, informed by the reality of racism on both sides of the veil, and use them to reimagine how racism is defined.

BEYOND DOUBLE CONSCIOUSNESS

Du Bois first becomes aware of the veil, a concept that is central to double consciousness, when he experiences discrimination and learns that there is a different set of rules, opportunities, and possibilities for Black people in the United States than there is for White people. The rules are more unyielding and unforgiving, the opportunities more rare and more difficult to come by, and possibilities stifled. For Du Bois, this knowledge is debilitating. It doesn't stop him from being productive—for certainly he was prolific—but it does spawn in him a double consciousness as he recognizes the internal conflict that results from living on both sides of the veil. The development of double consciousness, therefore, follows conscious or unconscious experiences with, or awareness of, the veil.

Remember that the veil is unilateral: Black people are able to see through it and understand the White world, but it does not work in the opposite direction.[18] White people, especially those who claim to be "colorblind," can be oblivious to the ways that race and racism affect the lived experiences of people of Color. Because of this, part of Du Bois's goal in writing *The Souls of Black Folks* was to lift the veil so that White people could see the results of racism, including pain, gross inequality, and an entrenched system of White supremacy that privileges White people and actively and passively combats Black efforts to rise. By lifting the veil through his writings, he hoped some Whites would recognize that the Negro struggle was not due to some innate or cultural inferiority, but to historical and contemporary experiences with discrimination at individual, group, and institutional levels.

One way Du Bois discusses being able to surpass the veil is through procuring an elite education that would allow Black folk to converse with White folk as intellectual equals in the languages of Western literature. I think we can problematize the idea that to be equal one must be intellectually elite. Our worth is certainly not tied to our access to education in a world that denies equal access for poor, minoritized peoples! Unfortunately, this is consistent with how Du Bois discusses intellectual elites elsewhere. The notion of the "talented tenth" is another example of how Du Bois believed the success of the Black freedom struggle was dependent on its most privileged persons.[19] The hypothetical solution to double consciousness, according to Du Bois, is dependent on individual performance, opportunity, and psychology.

Du Bois also talks about the veiled speech Black people often engage in in front of White people. Levine's research on Black "folk thought" explores the ways enslaved Black folk communicated with secret messages about and in front of White people without their knowledge.[20] For example, White masters would sometimes demand that the people they were enslaving sing songs for their entertainment, not knowing that the songs being sung had hidden meanings and were mocking them. Contemporary examples of this veiled speech can be seen in Black comedians earning a living from White people paying to hear jokes about Whiteness; hip hop artists whose music critiques Whiteness at structural or interpersonal levels but still resonates with

White audiences; or even Jordan Peele's film *Get Out,* a hugely successful mainstream horror film with clear statements about how White supremacy and anti-Black racism function in the twenty-first century.

This veiled speech can be food for the souls of Black folk who are able to laugh or cry with artists and prophets communicating about their collective experiences with racism. In this way, veiled speech may mitigate the burden of double consciousness. But veiled speech and slave songs differ from double-sided consciousness in that they are private, backstage, or encoded acts. They may represent communal strategies that can strengthen psychological resilience, but the hidden nature of these responses prevents them from challenging the reproduction of racism in mainstream spaces. Of course, when veiled speech comes from prominent figures, like Vince Staples or Kendrick Lamar, or is presented on a public stage, it can represent frontstage critiques of or challenges to Whiteness. These incidents are also distinct from double-sided consciousness, however, because the people are witnesses to but not partakers in these challenges.

Veiled reflection on the double consciousness conundrum, then, is typically reserved for spaces that are private or actions performed by individual actors who produce art for public consumption. When these discussions or reflections meant for the Black community take place through public-facing social media platforms, however, they change this dynamic. Double-sided consciousness can occur when Black people and other people of Color engage in communal discussions that are not meant for White consumption but, through technology, nevertheless become visible to White audiences. Double-sided consciousness, then, is veiled speech shared beyond the veil.

I'm tired of wypipo ☹

ever since i watched "Get Out" i been scared to go into wypipo houses.

Been listening to Good Kid Mad City a lot lately. It's gonna be a long time before I forgive wypipo for giving that Grammy to Macklemore.[21]

Double consciousness results from Blacks being forced to know the way White people feel about them. Black Twitter inverts this process, publicly and communally expressing how they feel about Whites, from disgust and exasperation to fear and resentment. Here double-sided consciousness represents a relinquishment of the collective need to ignore racism in an attempt to stay safe and employed in a White-dominated world, and disdain for the privileges associated with Whiteness.

AWARENESS OF RACISM

In chapter 5, I discuss the research on stereotype threat. The results are consistent: when individuals are primed with a stereotype related to themselves, that stereotype becomes a self-fulfilling prophecy; individuals are more likely to perform at a lower level on outcomes related to the primed stereotype. This is the research that best highlights why double consciousness is important (or problematic). The phenomenon Du Bois describes—constantly being vigilant in the face of perceived oppressive and hostile attitudes from Whites—damages the Black psyche in ways that may not be immediately apparent, but that bleed into our performance on a variety of outcomes and in a variety of contexts.

One dimension of double-sided consciousness is the rejection of the notion that Blackness is inferior and Whiteness is superior. The following tweets are from users who are not impressed by cultural repertoires they associate with Whiteness:

I love watching cooking videos but I just . . . wypipo really be fucking up the vibe

I knew wypipo didn't season their shit BUT Y'ALL NOT WASHING MEAT??????????????????? Yo [@cookingshow]

Ok are most wypipo just incapable of dancing on beat? Why do they fight against the beat so bad? It's not even that they can't dance, it's that their dancing is literally them struggling to be off beat.

Wypipo sex looks so trash. No rhythm, overly aggressive/borderline domestic abuse, unseasoned

There is an entire genre of internet videos that show folks of Color responding to White people cooking in critical and hilarious fashion. Nigel Ng is a Malaysian comedian who has built his brand on critiquing mostly White chefs who post videos of themselves cooking traditionally Asian cuisine. He even gave the famous chef Gordon Ramsey, known for his vehement critiques of contestants on his cooking shows, a taste of his own medicine. Similarly, the notion that White people can't dance, or do not have rhythm, is a commonly referenced trope that folks of all racial backgrounds have historically laughed at. For example, when Barack Obama was asked whether he, or Bill Clinton, was the first Black President, Obama responded by saying, "I'd have to investigate more . . . Bill's dancing abilities," to which a crowd of mostly White people laughed.[22]

In the third tweet, "harmless" cooking and dancing stereotypes (no seasoning and no rhythm) are taken to another level and used to critique White sex as seen through porn. Investigations into the porn industry have found that Black actors (especially Black women) are often mistreated and underpaid, and that search terms for Black actors can be offensive.[23] The suggestion that the dominant form of pornography features White, "unseasoned" sex represents a challenge to the power structures and dominant definitions of what constitutes "good" pornography.

These tweets do not eliminate the harmful effects of stereotypes. But they are examples of double-sided consciousness publicly boring holes into myths of White superiority, and folks of Color being eager to distance themselves from Whiteness, rather than attempting assimilation.

The Multidimensional Inventory of Black Identity is a survey scale designed to quantify different aspects of racial identity.[24] The scale includes measures of centrality, how important race is to one's sense of self; private regard, how one thinks about one's own racial group; and public regard, or how one perceives the world to think about their racial group. This last dimension of racial identity, public regard, is closely related to our discussion of double consciousness: it is a measure of the degree to which people of Color may be aware of racism.

Some research suggests that people of Color with low public regard (meaning they are aware that society looks down on their group) are less negatively impacted by experiences with discrimination.[25] And research on Black racial socialization finds that a common practice among Black parents is preparing their children for bias.[26] If young people of Color expect racism, they may externalize instead of internalize discriminatory experiences, and understand racism to be a problem located in White supremacy rather than in any personal shortcomings. For example:

> Yall so worried about what Wypipo must think of us that it makes you cringe to see a black woman being unapologetically herself in white spaces without being called out on it. Hate to see that.

Folks of color who are concerned with respectability politics may change their behavior in front of White people (or in private), including the way they dress, talk, or act, in order to be a good "representative" of the race and challenge negative stereotypes.[27] While this has been seen as a vehicle for advancement, it also may represent an outgrowth of double consciousness as Black people and other groups of Color spend inordinate amounts of energy seeking to control the ways that White people think about their racial group. In practice, this often looks like middle-class folks of Color policing the behaviors of poor folks of Color in order to prevent them from making the race look bad. Cathy Cohen's work on the politics of deviance critiques respectability politics and suggests that deviant choices (i.e., queer folks of Color rejecting White, middle class, heteronormative ways of being) have the potential to "open up a space where public defiance of the norms is seen as a possibility and an oppositional worldview develops."[28] Releasing ourselves from the preoccupation with what White folks think of us can be empowering.

This tweet highlights the effects of worrying about pleasing White people as it relates to comedian Tiffany Haddish, who was condemned by some Black folk for acting "ghetto" or unprofessional in front of the camera or at White awards shows. Double-sided consciousness takes a page from the book of disrespectability politics and encourages a shift from concern with the way folks of Color are seen by White folks, and towards critique of White practices and policies as a form of resistance.

Other research uses the term *racial battle fatigue* to refer to the net effect of long-term, regular exposure to minor racial slights or racial microaggressions.[29] I see racial microaggressions as being related to double consciousness because they are subtle racial statements that do not explicitly communicate racist attitudes, but implicitly communicate to people of Color their own inferior status. Given the inability of some White people to know when they or other Whites commit racial microaggressions, perceiving them can be thought of as a part of Du Bois's second sight: in contexts where overt racism is rare, for example, racial microaggressions may be the primary method through which Blacks and people of Color learn how Whites think about them (on an interpersonal level) and develop a double consciousness.

Just as the user-generated resistance we saw in the last chapter exposed microaggressive behaviors, some tweets on Wypipo discuss ostensibly innocuous White behaviors that are perceived to be connected to racism.

"I stumbled upon this" is wypipo code word for "I stole this from a woman of color, refuse to acknowledge her in any way, and I don't care."

For the 3rd time Octavia Spencer has been nominated for an Oscar for playing the role of a maid/mammy. I point this out not to downplay her tremendous talent, but it is telling that the roles written for her with the most depth and range are always in service to wypipo.

Leslie Kay Jones's research on Black intellectual production discusses the ways content from Black Twitter is often used "as fodder for mainstream media coverage of racialized events and as a means of tracking the popularity of ideas and products."[30] By publicly naming this plagiarism, Black women reclaim the right to their intellectual property.

Similarly, in the second tweet the user notes that an elite Black actor has been consistently given roles in which she is subservient to Whites, something that may be indicative of racial hierarchies in Hollywood and beyond. In both examples of double-sided consciousness, users publicly challenge the power dynamics that stifle the intellectual labor and creative talent of Black women. The world has heard these challenges. Tressie McMillan Cottom discusses how traditional media outlets have increased hiring among women of Color as a result: "By changing the tastes of tastemakers, Twitter made your media diet more diverse today than it was 10 years ago, with little effort on your part."[31]

FROM DOUBLE-SIDED CONSCIOUSNESS TO DIGITAL ACTIVISM

The problem of double consciousness is not resolved so long as it remains a private, individual-level struggle. For Du Bois, the solution comes not from a communal response to double consciousness, but an internal response to bring these two selves together. While individual-level solutions to double consciousness may exist, research on the individual-level effects of racism shed more light on the potential harm these types of experiences can have—or on the ways double consciousness can manifest in contemporary times—but do not necessarily provide directions for solving this dilemma. In fact, it seems that most writing on double consciousness accepts this state as static or permanent. Because double consciousness so acutely describes the way many folks of Color experience racism, in a way that (unfortunately) appears to be unavoidable, few if any scholars have attempted to solve the problem.

Double-sided consciousness describes alternative ways of responding to the forces that create double consciousness, and opens a realm of possibilities for resistance efforts that have transformative potential in changing how double consciousness is experienced. I suggest that using the term *Wypipo* not only creates and reinforces solidarity for Black people and all people of Color—people who are able to speak about Wypipo in the abstract, or from an outsider's perspective looking through the veil—but it also externalizes the problems of racism and Whiteness that are internalized in other contexts.

Cathy Cohen writes that "cumulative acts of individual agency are not the same as collective agency."[32] Unintentional resistance is not the same as organized activism. It can, however, redefine "the rules of normality that limit the dreams, emotions, and acts of most people."[33] This is what is happening with double-sided consciousness. Cohen's writing, which centers on Black queerness in the face of racism and homophobia, here illuminates the potential for collective discursive projects (like the usage of the term *Wypipo*) for reimagining the self-conception of people of Color in the face of White supremacy. Even though everyone who used the term *Wypipo* did not do so as a form of activism, the effect was resistance—a critical challenging of dominant norms and a reimagining of the social order. In this chapter, we see the way the usage of the term *Wypipo* served to invert double consciousness through the collective rejection of dominant narratives and the public sharing of counternarratives that problematized Whiteness.

Wypipo is far from the most prominent project or term that comes out of Black Twitter. I chose to investigate Wypipo in this chapter because the term is used in everyday Twitter conversation and is not a central part of any broader activist or organizing activity (e.g., hashtags such as #BlackLivesMatter). This is consistent with my broader agenda of understanding how technology shapes resistance amongst everyday folks of Color, and not just through the viral, global, and news-making modes that we have all become familiar with. Lots of research has explored these big-scale events and global hashtags and trends on Twitter, but fewer have explored the everyday resistance that we see here. Black Twitter is changing discursive strategies for disrupting racist narratives and behaviors in ways that are empowering for us all, whether or not we use Twitter.

In this chapter we see how some of the dynamics discovered on the college campus play out in a broader online setting: Black Twitter. Twitter can sometimes seem like its own self-contained universe. But from unmasking racism as users use the term *Wypipo* (to highlight problematic White behaviors and norms), to resisting racism through public-facing, digital cultural practices that engender new ways of thinking about the racial hierarchy, the digital practices in this chapter seem to mirror some of the processes we've been exploring throughout this book.

Will these changes be durable? To what extent do the processes described in this chapter carry over to the day-to-day lived realities of Black people and people of Color? Usage of the term *Wypipo* is not the only example of double-sided consciousness. As we see Black Twitter continue to transform the national conversation about race and racism, we should pay attention to the effect this may have on folks of Color and our changing ways of envisioning what it means to be free in an oppressive world—including both the ways online resistance can shift in-person power dynamics and strategies for resistance, and the limitations of online resistance and discursive projects that may not change the lived realities of people of Color unless they are connected to complementary activist coalitions and action for social change.

Beyond the narrative-based interventions discussed in this chapter, we're also seeing technology being used to harness, focus, and change what activism looks like in the twenty-fist century. In the next chapter I return to the college campus one last time to explore online activist efforts that have as much of an effect on the activists as they do their peers.

7

PROTEST, POSTERS, AND
QR CODES

The nonviolent approach does not immediately change the heart of the oppressor. It first does something to the hearts and souls of those committed to it. It gives them new self- respect; it calls up resources of strength and courage that they did not know they had. Finally it reaches the opponent and so stirs his conscience that reconciliation becomes a reality.
—MARTIN LUTHER KING JR., 1960, "Pilgrimage to Nonviolence"

DURING THE SUMMER of 2020 I took my children to their first protest. I found out about the event through social media: a caravan that began at the Cook County Jail in Chicago, then made its way to the South Side of the city. We were then less than six months into the first COVID-19 lockdown, but I felt comfortable taking them because we, and others, were going to be in the car (at least until I made a wrong turn, accidentally leaving the caravan: then we parked and marched outside with masks on).

We made "Black Lives Matter" signs to put in the car windows, and were followed by my sister, Reese, and brother-in-law, Danny, in the car behind us. My kids were already familiar with the Movement for Black Lives and why it was necessary, and were excited to chant and honk along with the hundreds of cars that inched forward around us.

Other cars had signs, too, which my kids were spotting and calling out to each other like a game on a long road trip. My oldest, Malachi, then an incoming eighth-grader, asked about one sign in particular that focused on prisoners in the Cook County Jail. "What do they mean, 'Release them all'?"

So, I explained to Malachi—and his younger sister, Karis, and brother, JD—what prison abolition was and the logic behind it. Prison abolitionists don't think the carceral system can be reformed; they believe it must be abolished.[1] We know that the prison-industrial complex is broken and biased. We know that jails don't work. So how can we envision a future with no prisons? I was in the middle of writing a paper on abolition, trying to articulate these ideas for an academic audience[2]. I didn't plan on teaching these concepts to children that day! But they learned by doing, watching their fellow protestors, and asking questions.

This protest had a very specific purpose. To bring attention to the injustice prisoners were facing in the face of COVID-19, and to create a community oversight committee for police in Chicago. Thousands of people gathered together to achieve these goals. But it also had an effect on the people who showed up. If you've been to a protest, you likely know the feeling of being in a crowd with a cause, bunched next to people who, like you, believe that there is something in the world that needs to change, and are spending their time and energy to push for that change to happen. The energy can be electric.

But beyond the powerful sense of being surrounded by people who are collectively demanding a better world, my children learned more than I expected them to learn. As their dad, I don't have a curriculum or plan for when to introduce them to justice concepts. Perhaps this is wrong, and I should be more intentional. But I just seek to maintain a positive, open relationship with them, so that they are comfortable voicing their observations, questions, and concerns with me. Then, when questions come up, I answer them to the best of my ability, and honestly.

That day, their learning was impacted by the abolitionist message that someone—I'll never know who—took the time to write on a posterboard. This is the unintended consequence of activism: when you get involved, you can learn, grow, and sometimes become someone new. I think the same thing can happen with online activism. Activist and justice-oriented content posted online—in the form of articles, blogs, videos, comments, or tweets—can have the same effect as that abolitionist poster—markers on paper, taped to a car window—and help other internet users learn and grow in their understandings of race and racism. My interviews with students showed that many of them felt that much of their own learning and growth around issues of race and racism took place online. For example, Amanda talks about her time on Tumblr, a blogging site, saying,

> I was following these people who were like, "racism is like not just someone calling you nigga it's, like, . . . Black men being in prison disproportionately" . . . it was just this eye-opening thing for me. 'Cause I came in very, like, "Stacy Dash" . . . and now I've left and now I'm just very, like, "fight the power."

Stacy Dash is a Black and Latinx actress, former Fox News Correspondent, and self-described "angry Black conservative lady," who is well-known for making offensive anti-Black remarks.[3] In 2022 Stacy Dash apologized for some of her offensive comments, but Black Twitter mostly made jokes at her attempt to become part of the Black community again, something they saw as being a result of her unemployment, not a change of heart.

Amanda had an affluent upbringing, said she had never personally experienced racism before coming to college, and talked about feeling at times like she had more in common with her White friends because of class differences between her and many of her peers of Color. For example, she mentioned that her White friends were more able to afford to eat at the type of restaurants that she enjoyed frequenting.

Perhaps the reason Amanda believed she had not experienced racism before coming to college was that her understanding of racism was limited to its more overt forms. The blogs she followed unmasked the ways racism operates on a systemic level or can be implicit in friendly interactions. Of course, Amanda experienced in-your-face racism on campus, too, like the time she was called the n-word at a lunch table.[4] But remember that Amanda saw this event as being one of many microaggressions, indicating that at that point in her development she was aware of and bothered by more subtle forms of racism as well.

For Amanda to liken herself to Stacy Dash indicates that just a few years before we talked, at best she had a rather limited understanding of racism, and at worst may have internalized racist narratives and attitudes towards Black people, especially those with fewer resources than her. Conservative ideologies (like those espoused by Dash) would attribute her family's class advantages to hard work and discipline, and the failure of other Blacks to reach a higher economic status to personal, moral, or cultural failings. Now Amanda describes herself as militant, which typically refers to far-left approaches to pro-Blackness and antiracist thought, including the understanding that racial and class inequality is due to systemic oppression, not broken culture or a lack of effort.

When conservative pundits began the uproar against critical race theory, and politicians worked to ban the theory from schools, I remember wondering, how many K–12 programs actually include CRT as part of the curriculum?[5] It's not something I'd ever heard of being used with school-aged children, and is more likely to be found in advanced college or grad school courses on race. For students like Amanda, critical awareness of racism tends to come from informal online sources, not textbooks.

Other students talked about how online sources helped them understand their own racial and ethnic identities. Judith said that she didn't consider herself Latina before finding Tumblr blogs with a focus on identity that would "ask their followers to submit selfies of themselves and just kind of celebrate the diversity of them. So I participated in that." For both Amanda and Judith, the blog posts they found were like abolitionist signs in the window, affirmations for women of Color that did more than make them feel good—they may have prepared them to be more resilient in a world that does not believe them to be as worthy of admiration as do the blog and its followers.

For other students, signs in the window encourage them to think about the world beyond their own racial and ethnic identities, and learn to empathize with, and potentially fight for, other communities of Color that are in the struggle against White supremacy. For example, Jordana, a Black student in New York, says,

> I am Black and that's the experience that I have. But I think sometimes it's a wake-up call to see certain discriminations that happen to people that are, like, Hispanic or . . . Middle Eastern, and just seeing what they deal with . . . some of it is almost as extreme as what the Black community faces.

For folks of Color, seeing others discuss and perform race on social media can confirm ways of thinking about the world that they were unsure about, or give them language to describe things they had experienced. Reba, a Latinx student in Chicago, told me about watching a video that discussed Miley Cyrus's cultural appropriation of Black dancing styles, and said it gave her words to describe the discomfort she had been feeling when watching Cyrus videos. And after watching a YouTube video explaining the strong Black woman stereotype, Sara realized that the stereotype explained her White friends remarking on her "strength" despite her not doing or saying anything that she felt was strong. This pattern of awakening taking place due to online racial content was consistent in my interviews.

Natalia, a Latinx student in LA, echoed this way of viewing content on social media as a wake-up call, something that exposes you to things that you don't see in "your daily routine," and even suggested that the internet is where "people that have social media in general get most of their information or their ideas of what race is." I imagine the difference between one's "daily routine" and explorations of race and racism on social media and online might be even more pronounced for some White folks.

For example, while I, too, am taken aback when I see videos of White women calling the police on Black people for no reason, or videos showing anti-Black police violence, I am not surprised by them. I have had too many personal experiences with the police pulling me over for no reason (or because "there have been robberies in the area") or with White folks who think they can police my behavior in public spaces to be surprised by these videos. I know too many people who have been physically hurt by the police, or have faced ugly, hurtful, and embarrassing types of racism, for video evidence of these behaviors to astound me.

But for a White person who exists in a social context (real or imagined) where the racial niceties are observed, those videos can be *proof* that the world is not always a *nice* place for people of Color. Like the PIC in chapter 4, they may very well unmask racism and surprise them with a face they did not expect. My colleague Courtney Cogburn developed a virtual-reality scenario that allows users to put on a headset and experience a day chock-full of racism from the perspective of a Black man.[6] This work allowing users a chance to embody the Black experience in 360 degrees has been extremely well received.

While social media is not as immersive as virtual reality, it may have some of the same effect, serving as a painful exhibition of how race continues to dictate the health and happiness of people of Color. Plus, social media is interactive, allowing users to connect with other users who are responsible for sharing some of these counternarratives. For example, Jeff, a South Asian student in Los Angeles, said, "You hear about all this conflict going on in Palestine and back in the day we didn't have any means of connecting to them, it's like so far away. But with the internet you can see visuals coming from the Middle East . . . we get a different perspective."

Jeff identifies as an activist, and for him the videos from Palestine were an entry point to making personal connections with activists across the globe. In this way, social media posts about racism or anti-racist activism can be more than just messages in the window of a moving car—they are like metaphorical signs in the window that also include a QR code.[7] We scan the code to open up another page with more information. We are given directions to follow the car, are able to ask questions about the poster's message, and receive notifications when the driver makes new posters. There is even a place on the poster where we can add our reactions or comments.

The person who wrote that simple abolitionist message on their car will likely never know its effect on three young people and their father. But as Jeff's example indicates, many online activists hear from the people they influence. For an activist, knowing how messages are being received by tracking views or shares and engaging with com-ments or messages can be confirmation that they are impacting the world in the way they intend.

Along these lines, Cristina tells us about the feedback she receives and reputation she built for consistently posting justice-oriented con-tent online, saying:

> People like make fun of me about all the time. They're just like, "Cristina gets like 5 billion likes on anything that she posts." You know? "Cristina's, like, the most popular, blah blah blah" . . . I'm one of those people who, like, annoyingly, all the time is talking about these kinds of things. And there are more people like that. You know? As I do it more, other people do it more . . . There's just been, like, a very sea change in the dialogue that we are having amongst each other as students about issues like race . . .

Cristina also shared that her circle of activist friends influences her thinking as much as she hopes to influence others, saying, "my newsfeed is like hella liberal, hella queer, feminist, like, socially aware . . . they are constantly providing me new things to think about." If we can return to the caravan metaphor for a second, Cristina makes signs with a group of friends who carpool to the caravan. Cars honk to let her know they appreciate the signs in her window, and she begins to see other cars join the caravan with similar signs in their windows.

The likes on her posts are, for Cristina, evidence that the activist work she engages in is having the desired effect on her peers and her campus community. As she continues to post, she recognizes that people in her circle do the same, saying, "If this momentum keeps . . . ten years from now we're just going to be a much more sensitive campus." This is a long-term goal for Cristina, to use her activism to change how race is discussed and experienced on campus, even if the change is not fully realized during her time as a student.

In 2020, the sustained protests and uprisings in response to the murders of George Floyd, Breonna Taylor, and Ahmaud Arbery shook the world. This of course included more people hitting the streets to protest than have ever been seen (something I'll discuss a bit more in the next chapter), but also followed nearly a decade of online activism and organizing, as activists, antiracists, and hashtags like #BlackLivesMatter, #Ferguson, and #SayHerName brought issues of anti-Black police and vigilante violence to the forefront of national consciousness.[8]

The goal of unmasking racism, of challenging racism on campus and racist narratives and actions around the globe, is to dismantle White supremacy. I think it's important to understand the transformative effects the type of resistance we have been exploring in previous chapters, from online racial checking to double-sided consciousness, can have on racial, social, and political consciousness development. In this chapter I explore digital posters in the window—or online activism—and talk with students about how they engage in, and are affected by, activism both in person and online, and whether these are as separate as we might think. Internet technologies have had a nearly incalculable impact on activism and organizing around the world in the past few decades. While some organizers still hold to the old ways, simply using social media as a tool to communicate with their communities (a more modern mail or email list), in other cases we have seen internet-based activism morph into something else entirely and force us to think differently about organizing.

What are the opportunities and limitations associated with traditional, in-person forms of organizing, on one hand, and online activism, on the other? Can they be used in tandem? Should we consider the type of consciousness-building activities that help people learn to think deeply about race and racism activism, or is it something else?

ORGANIZING FOR CHANGE

At this point in the book, I've used several examples from the civil rights movement because in my view, there is no better way to learn about activism. Most people are familiar with the movement, but many also misunderstand it. Every February, during Black History Month, children in America learn about Martin Luther King's most famous speeches, that Rosa Parks was tired and sat down on a bus, and Black people and their allies braved dogs, firehoses, and police sticks on their march to victory against racism—the Civil Rights Act of 1964. The framing of the movement as being powered by the "dream" of one man, or that a woman like Rosa Parks only contributed to the movement by accident (i.e., just because she happened to be tired), belies the realities of the movement as being built on decades of organizing, much of it done by Black women who do not make the history books (or at least, the textbook versions—many critical historians tell the real story—see Payne's "Bibliographic Essay" for an extensive review and discussion of critical and uncritical writings about the movement).[9]

If we believe the dominant narrative about the movement, we may think we need another charismatic leader like Dr. King to continue the fight. But if we realize that the movement was shaped and fueled by the dedication and persistence of everyday people, we might be convinced that working together we can make a difference and pull down oppressive racist structures that still shape our world. Highlighting the parts of the movement that do not make the history textbooks can illuminate the types of resistance that are most important for continuing the struggle against White supremacy. Charles Payne quotes Ella Baker as saying, "I have always felt it was a handicap for oppressed people to depend so largely on a leader, because unfortunately in our culture, the charismatic leader usually becomes a leader because he has found a spot in the public limelight. It usually means that the media made him," and, "Strong people don't need strong leaders."[10] We do not need a charismatic leader to carry us to the Promised Land. Collectively, and with strategic organizing, we can find and fight our way to freedom.

Resistance against racism during the civil rights movement can be placed into two broad categories: mobilizing and organizing.[11] Mobilizing refers to creating short-term events that are designed to maximize impact and media attention. This is exemplified by the work of Dr. Martin Luther King, who would travel from city to city, leading front-facing resistance efforts with speeches that live on in national memory in myriad forms, from fiery video clips to poetic excerpts in textbooks. Organizing, on the other hand, refers to a long-term commitment to developing leadership. This style of activism is less flashy. When there are no cameras present, when the world isn't watching, organizers spend years recruiting and training activists, building expertise among everyday people.

Of course, mobilizing and organizing are often interconnected, with the most successful mobilizing being preceded by much organizing work. The Montgomery Bus Boycott is an example of a mobilizing event which had a large impact in a short amount of time. But there was much organizing behind the scenes—from strategically establishing Rosa Parks as the face of the event who would spark the protest by refusing to give up her seat (her refusal to move wasn't because she was tired—it was planned and intentional!), to making sure boycotters had rides to work—that made the event possible. Ella Baker is perhaps the most prominent civil rights–era organizer, whose decades-long work resulted in the training of generations of activists. The highly publicized mobilizing efforts of the civil rights movement, like the Montgomery Bus Boycott or the March from Selma, were possible because they had been preceded by decades of organizing work that developed the requisite leadership and organizational infrastructures that could support such large-scale mobilizing efforts. This is central to understanding what Jacquelyn Dowd Hall calls the long civil rights movement, which conceives of the movement beginning in the 1940s or earlier, not 1954, and extending through the 1970s, 1980s, and, some argue, even to the contemporary Movement for Black Lives.[12]

The types of activism we see today with the Movement for Black Lives often falls into these two camps as well. Many BLM activists do the hard work of on-the-ground organizing (both face-to-face and online) that enables the large-scale events we saw in 2020 and will continue to see in the years to come. Mobilizing and organizing are difficult tasks, and for organizers especially, often the work can entail years of labor that appear to be fruitless.

There are different schools of thought within organizing, each with distinct visions of social change, methods of activism, and undergirding philosophies.[13] The most well-known is the power-based model, or the Alinsky model, which uses aggressive tactics to demand a seat at the political table. This type of organizing focuses on building community capacity to engage in direct action or protest. The community-building model focuses on strengthening communities by building the capacity of residents to shape their communities. The civic model focuses on strengthening formal and informal social controls as a way of preventing the problems associated with social disorganization. The women-centered model prioritizes the caretaking of families and children in the community. And the transformative model emphasizes addressing structural-level problems that can only be solved with large-scale social change. Activists using the transformative model focus on training people to question dominant frameworks and providing an alternate vision for society geared around justice.

Daveena has been involved in organizing in response to a number of racist events on campus. She serves in leadership in cultural organizations and is part of several student-administrator groups geared towards improving the campus racial climate. She is one of the only people I talked with who identifies strongly as an activist but eschews online activism, choosing instead to focus solely on in-person forms of organizing. While her perspective is not prevalent in the data, it provides some insight into some of the perceived benefits and limitations of in-person activist work on campus.

For example, a number of students in Chicago spoke about an incident when a campus fraternity planned a racist-themed party and posted about it on Facebook. Daveena was one of the students who organized in response to this event. She was part of a team that wrote letters to the administration, met with a group of administrators, and, after soliciting an official apology from the fraternity, was on a panel with a fraternity leader to discuss the incident.[14] Daveena notes her disappointment with her experience on the public panel.

> We had, like, three people on the panel who were Latino . . . a lot of people who were there in support of Greek life ended up asking a lot of questions that were attacking the Latino community and the people of color on campus. Pointing out "reverse racism" . . . There was one guy . . . he asked a question at the end demanding to know what we had all learned. Like, "we heard what the frat people learned . . . but what have YOU learned from this experience?"

This example highlights some of the strengths and weaknesses of traditional forms of organizing on the college campus. First and foremost, it was effective. Daveena and her partners were able to use traditional organizing methods to get school leadership to pay attention and uphold the second rule of racial discourse: limit race talk to the niceties. This was no small task. As we discuss in chapter 3, White folks are often given exceptions to these rules, but in this case the fraternity was made to apologize and change the racist party theme. Because they had "won," Daveena, and other Latinx students assumed the purpose of the panel and event was for healing so that both groups could move forward together as peers. Instead, the questions asked at the panel reminded Daveena of her marginal social position on campus, despite her successful organizing efforts. Rather than provide healing, the event reinforced racial and gender norms and power dynamics on campus. The fraternity had its hand slapped, but members and supporters at the event made it clear that their thinking about race had not changed.

In addition to this disappointing conclusion, Daveena talks about the toll her work took on her academics.

> Administrators warned me . . . "You don't want to get back-taxed on your homework" . . . I look back on that time and I'm kind of frustrated with myself . . . if I had known how the rest of my time at Chicago . . . would go, I would've taken more time to just hit the books. But I didn't because I thought it was more important for me to work on issues that would affect my community as a student and as a Latina. Because ultimately, if I wasn't going to do anything, then no one was.

The leadership and initiative Daveena demonstrated with this organizing shows how promising of a student she was and could translate to success in any number of jobs or graduate-school programs. But if her grades suffered, this might limit her future options. Daveena regrets her grades dropping as a result of her activism, but she believed there was no one else to do it. When I interviewed her, it was over a year after this incident and she had "retired" from activism, laughing about how the current "incidents" on campus were the responsibility of the younger students, not her, then an upperclassman.

This story reveals some of the limitations of traditional, in-person forms of activism on campus. For Daveena, the cost was high and the payoff was low. She felt alone, and this lack of support meant she felt responsibility to give more time and energy than she could afford as a student. In chapter 5, I discuss how the responsibility to respond to microaggressions or other forms of racism is lessened in online counter-spaces, because people have increased and immediate access to more support and allies. Similarly, double-sided consciousness is a collective undertaking. Online activism may be less taxing given this increased support and decreased perception that if one student does nothing, nothing will be done.

Daveena's negative experiences with the student forum also don't seem to be unique. For example, Peyton, a Black student in New York, talks about forums for racial discussions on campus.

> I don't go to community columns or the big conversations like that because they're so exhausting. They're all the same . . . there's usually one really, really guilty White person. And there's another person who's, like, "we used to not be able to drink from the same water fountain, racism is over." (laughing) . . . Then there's like, White women who try to relate . . . "I totally get it 'cause I'm a woman."

Forums like these often include a guest speaker or a panel around a topic related to race. Sometimes they are planned in response to some adverse event that took place on campus or demanded by student activists, but they also can be part of a university's attempt at creating a more inclusive racial climate. In theory, they should be places for students to heal, learn, make connections, and ultimately have a positive impact on campus climate. Instead, for both Daveena and Peyton, the racial power dynamics at the events mirrored the rest of campus life. Peyton describes attendees demonstrating White guilt, minimizing racism, and comparing race to other forms of disadvantage (gender or sexual orientation) in a way that centers Whiteness in a space that was meant to focus on the experiences of folks of Color. In Zeus Leonardo and Ronald Porter's work on "safe spaces" for racial dialogues, they argue that these spaces are only safe for White folks and privilege White comfort over honest discourse, which can (and perhaps should) be uncomfortable as participants engage with the painful realities of racism.[15]

Other students had different experiences with in-person forums. Valerie thought they were productive when they were populated by "people who wanna be there and people who wanna talk about it." But when forums are mostly attended by students of Color and White allies, other students felt they were preaching to the choir. For example, Dolores says, "It doesn't really help if the people that you're trying to target don't attend these things."

Hearing Dolores and Valerie's opposing perspectives on the makeup of in-person forums and events, and how both of their experiences differ from Daveena's, demonstrates the complexities of organizing antiracist spaces on campus. When spaces are majority White, and traditional rules of racial engagement apply, the event might feel disempowering for folks of Color. But if only people of Color and their allies attend, activist students may feel like the event is not really making a difference. This contradiction highlights the reasons online counterspaces are so unique, and how online activism might overcome some of the limitations by allowing folks to get involved on their own time, and in racially integrated spaces where the racial power dynamics do not feel quite so oppressive.

Adding even more complexity, Peyton and other students in New York talked about conservative speakers—whose rhetoric they believed to be racist—being invited to speak on campus by conservative student groups. They were bothered by university resources going towards these speakers and noted with appreciation that many liberal White students would often respond to these events with protest. Peyton was not a part of the protest, but got close enough to get a good look and recalled a friend saying to her, "Did you notice not a single brother was in that protest?" In response, she said,

> I guess that's because we have class. And he's not the first person to have said this, so I'm gonna go to class. Are they gonna kick me out because of what this [man] said? No! Am I going to get a worse grade if I don't go to class? Yes! So then I'm going to go to class. But all of these White people were, like . . . "He said that!" And I was, like, "The chief justice of the supreme court said that." People say it. If I were to protest everyone who ever [said it] I would never go to class. So I'm going to go to class.

Peyton's decision to focus on her work may indicate a difference in priorities, not just in terms of attention to schoolwork but also in the types of racism she thinks are worth fighting, versus those that White students at the protest think are worth fighting. Racism for many students of Color is more than just a theoretical or symbolic threat. It has an unavoidable and damaging presence in their lives. But for White students who identify as antiracist yet are not harmed by racism, highly visible racist symbols may be more loathsome than the invisible machinations of racism that undergird White privilege (and negatively affect the experiences of their peers of Color). This is not because of ill will. When a speaker breaks the rules of racial engagement, like this conservative speaker saying things that were interpreted as racist, racism is unmasked and easier to both recognize and challenge. Masked racism, like laws, policies, or even microaggressions on campus, are more difficult to identify and protest, but may have a more consistent effect on the experiences of students of Color on campus.

In this case, Peyton anticipated the academic cost of activism and decided it was too high (something Daveena might have wanted to do in hindsight). Peyton believed the speaker's racism was "business as usual" in many contexts, including, she notes, the highest levels of the justice system. Deciding when to respond to racism is difficult for people of Color who recognize the omnipresence of racism in a racial-capitalist society. No one has the strength to fight all the time.

There are of course many students of Color who are very interested in attacking symbols of racism. Language and symbols have consequences, even when they may not be immediate or direct. For example, when Trump went off-script to describe COVID-19 as the "Chinese virus" in 2020, many worried about what this legitimation of a false racist narrative would have on Asian folks in the United States and worldwide.[16] We cannot separate the dramatic increase in anti-Asian hate crimes in 2020 and 2021 from Trump's harmful language.

On college campuses, student perceptions of whether in-person activism will make a difference may be driven by the level of trust they have for their peers and their school. Activism is difficult, and it has a cost, whether physical, emotional, or academic, that not everyone can pay. Understanding the costs of activism, and the frustration with not seeing effective examples of on-campus activism that lead to real change, may explain how for some students of Color online activism may be a more attractive form of engagement with racial-justice issues.

THE NEW SCHOOL: DIGITAL ACTIVISM

Social media has transformed the public face of activism around the world. From the Arab Spring to the #MeToo movement to the Movement for Black Lives, internet-based activism is changing how we participate in protest and resistance. Most of the time when we're talking about online activism, we're focusing on big-ticket items: the types of large-scale events that demand national media attention, bring sweeping changes to an industry, or are accompanied by in-person protests. Sarah Jackson, Moya Bailey, and Brooke Foucault Welles explore some of the most prominent examples of hashtag activism over the past decade, including #MeToo, #SayHerName, and #BlackLivesMatter.[17]

They find that marginalized groups and their allies use online networks to create counterpublics where they challenge racism, sexism, homophobia and transphobia, and spread counternarratives that empower marginalized communities. In the last chapter, I discuss Ray and colleagues' idea that hashtags form collective identities and how this might relate to double-sided consciousness.[18] Even when hashtag activism doesn't target a specific policy, it has an impact on the ways we think about ourselves and the world, spreading liberating ideas and concepts that originate from scholarship or activism to mainstream audiences.

But social media is also renegotiating what it means to be an activist, even when tweets or posts don't go viral, or when people don't use hashtags at all. Twitter has been the most important social media platform for the big-ticket activist events, and research on online activism rightly focuses on Twitter. But less than a quarter of Americans are on Twitter, and as I mention in chapter 1, only a few of the students I interviewed talked about having Twitter accounts. When they did, their Twitter networks were often separate from their campus networks. Through hashtags, young people can see and join activist work being done all around the world. But more often, they may be performing activism in their personal networks, independent of hashtags or national or international online activist efforts.

Young people are using social media to learn about politics and justice, sharing these things with their friends, and finding ways to get involved from their cell phones. Cathy Cohen and James Kahne call these behaviors participatory politics, defined as "peer-based acts through which individuals and groups seek to exert both voice and influence on issues of public concern."[19] This can include learning about political issues through online sources, sharing information in online networks, producing politically themed content (including blog posts, pictures, or videos), or mobilizing for real-world political action.[20]

Cohen was one of my mentors in graduate school, helping to shape my understanding of online resistance (and even helping me find the right traveling soccer club for my kids). Beyond her academic work, Cohen has supported the development of young tech-savvy activist groups like the BYP 100, though she takes no credit and they exist independently of her or her lab. Cohen's research finds that while some old-school political scientists have suggested participatory politics is not a serious form of civic engagement, it actually predicts increased political involvement once teens turn eighteen.[21]

An experimental study of 61 million Facebook users found that those who were shown a message at the top of their Facebook pages about voting were significantly more likely to vote when the message was accompanied by the faces of several Facebook friends who had already reported voting.[22] While online political messages are influential, they become even more so when they come from actual associates than when they come from advertisements or other generic sources. It may be that the effectiveness of participatory politics is related to its social dimension.

Online tools can lower the barriers to civic engagement and activism, but not everyone believes that internet technologies improve activist efforts. Some insist that online activism, or "slactivism," is a weak, watered-down form of activism with little potential to create real change.[23] From this perspective, organizers should focus less on digital strategies, and more on real-world, physical organizing that centers humans and not followers. Critics of online activism seem to think that this is a zero-sum game, and that the rise of online activism portends the end of traditional organizing methods. But this isn't true. Not only can participatory politics be a pathway to traditional politics—such as voting—but online organizing has consistently been tied to sustained in-person protests, especially with the Movement for Black Lives.[24]

The research coming out of the participatory-politics framework thinks about outcomes in terms of political action. To what extent does the sharing of information make young people more engaged? But formal political involvement isn't the only outcome that matters here.

Activism can affect how we think, or how we understand and learn about issues of racism and injustice. Online activists do not limit their measures of success to rates of voting or protesting. Instead, racial, political, and social consciousness are considered outcomes as worthy of attention as policy change. For some online activists, as we see below, the goal is to produce young people who think critically about race. If they adopt a critical and intersectional framework for understanding race, this framework is expected to influence how they think about, respond to, and fight for issues of racial justice down the line.

One of the qualities that seems to be consistent among activists is the feeling of responsibility for helping raise consciousness amongst their peers. For example, Rebecca says,

> You never *have* to educate. It's never your responsibility to do that. But, if you take it upon yourself, if it's something that you want to do, you have to think about how you can do it . . . How do you make it so that people are okay with being uncomfortable? You have to be uncomfortable. It has to happen. You're not going to progress unless you are made to feel uncomfortable. I believe that.

Increased access to support and decreased pressure to respond to racism are two important characteristics of the types of online resistance we've been exploring in this book. The online community of support, full of people who may themselves decide to respond to problematic situations, is uplifting for folks of Color who do not feel they have the time, energy, or expertise to respond. Given this dynamic, the decision to make injustice your responsibility—making the time—may be a part of what makes someone an activist. With online tools, activist students do not have to wait for administrators to show up or respond to their emails. They can speak into the situation right away. Rebecca's interventions are strategic, anticipating that critical lessons around race may lead to White folks, straight folks, or men of Color feeling guilty or defensive. She is therefore intentional about framing her interventions in ways that encourage them to stay in the conversation even when it becomes uncomfortable, thereby breaking the fourth rule of racial discourse from chapter 3.

I've been teaching race for a decade, and I still work hard to figure out how to do what Rebecca is describing. The burden to educate, as Rebecca puts it, should not be on students of Color. But they are doing the work anyway, and sometimes it seems they do it better than the professionals. Especially (and unfortunately) when they feel the work is not being done by their schools or institutions. Camille is a Latinx student in Los Angeles who also feels it is her responsibility to post about race. She says, "I kind of felt, like, sort of a responsibility, like I need to post about it, I need to let people know because if I barely found out then imagine they probably don't know."

Race-conscious student activists step up to fill the gap when it becomes apparent that many of their peers do not have access to critical frameworks for understanding race or counternarratives that debunk stereotypes or dominant ways of explaining away racial oppression. Camille talked about people following her on social media after seeing some of her posts, and goes on to say this about how she thinks of herself as an activist: "It might seem impossible, but I want to be an activist, like that's what I want to do. Because all of this really, really bugs me . . . I've been posting a lot of police brutality against [people of Color]. I feel like the more I keep posting about this the more other people will be like, 'hey, you know, this really does matter.'"

Camille goes on to talk about a specific instance where she kept posting about a police shooting that her friends and the news media seemed to forget about. She felt it was her duty to remind people that the officer still had not faced any consequences. We have seen a few police officers held accountable for violence against Black folks, including Derek Chauvin being convicted for the murder of George Floyd. But we must remember that convictions are rare, and getting away with violence is still the norm for police. The City of Chicago paid out at least $67 million in police misconduct settlements in 2021 alone.[25] Some of these situations made the news, but many more are situations where people were hurt, and then received settlements, without the world ever knowing about the situation.[26] Camille's actions in this case mirror the strategies hashtag activists might use: making an incident trend and putting pressure on police departments to respond.

Another feature of online activism is acting with purpose and intentionality. Examples of participatory politics such as sharing information online can seem like things people might do without identifying as an activist, but in talking with Jeremiah the difference between his activist work and casual postings about race online are clear. He says,

> They think that people talk about race to talk about race. They think that it's a bunch of victim stuff. They think that it's a bunch of us, you know, us being activists—people of Color and other people who care about these types of issues making it an issue, rather than it actually being a very real issue. And so, like, all my work is about showing people how it is an issue.

Jeremiah is involved with online and in-person social action, and in the months following our interview helped organize a campus-wide, week-long event focused around helping people understand how their social positions (i.e., their race, gender, orientation, or ability) impact their lived experiences. His activist identity goes beyond his online presence. But while in-person organizing events are less frequent, he takes his daily online conversations very seriously as sites of resistance. By giving people the intellectual tools to critique racist structures, ideologies, and actions, he enacts social change, helping people in his network to increase their racial and social consciousness. The positive feedback he receives about these actions not only reinforces his identity as an activist, but also gives him incentive to continue these types of discussions. He notes:

I'll get a couple of different Facebook posts on my wall that are like, "Man, you always have these discussions. It's really good to read these." Also, I get a lot of private messages and that's what I really enjoy and that's why I keep doing it. I've gotten dozens of private messages that say, "Continue to do what you're doing," because they're reading these posts, seeing my responses and seeing other people's responses . . . like, "I had this opinion and now that I see the type of dialogues that are happening, I see what you're saying in a way different light than where I'm normally predisposed to believe."

Participatory politics can be casual. Some people share articles or podcasts without intending to do activism, just because they find those things interesting. But for activists, those activities have a purpose. For Jeremiah, the focus of his online activism is ideological transformation, and the proof that his activism is working comes from the feedback he receives from his Facebook friends. Sharing a link on Facebook may seem as simple as walking down the street. But when you walk down the street with a sign about abolition, people may begin to walk alongside you.

Destiny is a student who engages in online actions that look like activism, but hesitates to call it activism because she does not feel she has enough followers (or hasn't received enough positive feedback) to claim the title. She says,

Once my name is a little more known, I will actually want to do activism through that. And now the best way for me to do so is here, going to [cultural group] meetings and trying to do that as an admissions ambassador . . . but online I don't talk a lot about race, simply because if I do, I know I'll get backlash, and not a lot of supports, because I've—like, I'm not a name, you know?

On mainstream media, you often need a "name" to be heard. Mainstream news outlets may limit their invites to the most prominent scholars, activists, influencers, or media personalities. But on social media, anyone can make a post about a social issue—this is part of the benefit of online organizing. Activists are able to bypass traditional gatekeepers of information, from mainstream institutions to mainstream news sources.[27] You don't need a PhD or to have written a book in order to be heard.

Destiny's perspective here, therefore, is surprising. She was the only person I talked to who felt she had to build a reputation through in-person activism before she had the credibility necessary to become an online activist. This, to be sure, is how some very prominent online activists have gotten their start, working with activist organizations and slowly building up their online presences until their activism appears to live more online than offline. But it is not typically how students discussed doing the work.

There is also a contradiction in this case, because in her interview Destiny talks about doing some of the work that we would consider activism, challenging racism both online and in-person, referring to these small challenges as knocking down dominos, saying, "Every domino that falls is just furthering the line of the falling dominos." One of the strengths of online activism is that there are lower barriers to entry. But if positive feedback, likes, or followers are perceived to be necessary to do the work, might some young activists be scared to engage? This is a potential unanticipated limitation of online activism. If young folks think they need to go viral, like hashtag activism, in order to join the Movement, this belief runs counters to the benefits of online activism being more democratic and egalitarian.

Unfortunately, just as some students receive positive feedback encouraging their activism, others receive implicit messages that these types of messages are unwanted. For example, Peyton talks about a less than stellar response to her postings about race.

> I've noticed that as of late my Snap stories have become a lot about race, and being that most of my friends are White, the views go down. Like,

you see the first Snap and its like forty-five people, and then the second one is, like, thirteen people . . . They usually don't get past the first one.

In Snapchat and Instagram story posts, users can post a series of images or videos, each a few seconds long, and their friends and followers can watch them in succession. Both apps, however, give users the chance to skip the rest of one person's story and move on to the next set of videos on their friend list. If Peyton is concerned with keeping her friends' attention, this type of subtle negative feedback—noticing that her friends skip her Snapchat posts when they are about race—can be discouraging. If positive feedback is part of what helps students begin to identify as activists, can negative feedback discourage activism?

Researchers use the term *echo chamber* to describe the way many online communities are populated by people who have the same opinion as one another on a certain topic.[28] But college students tend to be social media friends with large numbers of people around campus, not just people who agree with their racial and political beliefs. One of the benefits of online resistance like online counterspaces or social media-based activism, in fact, is that it reaches a broader audience than do in-person protests, forums, or campus student groups. Instead of seeing the same old faces showing up to every event, online you may engage with a different crowd, people who do not attend events in person but can read and respond to posts. But the type of response Peyton describes demonstrates that digital tools, like muting or skipping posts from friends who disagree with you, may further the creation of online echo chambers even among college students.

In other situations, online activism can be intentionally directed at a particular group. For example, Rachel talks about the permanent online counterspace she created, saying, "I started a group called [The Sister Space][29], and we usually discuss race issues, and particularly race issues related to women of Color, or gender issues related to women of Color."

Women of Color often feel marginalized in nonintersectional conversations about either racism or sexism, and therefore may benefit from spaces where theirs are the only voices. As an activist, Rachel facilitates the conversations in order to provide a much-needed space for women of color to grow and process their experiences. There are numerous examples of these types of groups, spaces, and social media accounts online. I follow several on my personal social media accounts and benefit from how critical and current their postings tend to be.

LOOKING BEYOND THE CAMPUS

Division I student athletes generate millions of dollars for their home institutions but are not paid for their time beyond basic scholarship plans. Some have argued that there should be some profit sharing among college athletes in order to take less advantage of these young people who put their bodies on the line.[30] In a similar fashion, in the absence of effective programming or educational opportunities around issues of race, some students of Color give their time and energy to educating people about race, resisting dominant and oppressive notions about what race means, and engaging in critical and intentional informal education efforts online. These students are activists who are helping to stimulate increased racial consciousness among their peers and are shaping the futures of their schools.

Many online activists seem to have a broad conception of activism. When they aren't pushing for a specific policy change, they still push a certain critical frame of thinking about race, racism, and inequality. They unmask racism and critique racism on various scales, from situations that make the national news to the interaction they just had in the grocery store. In the process they teach their followers what racism looks like, how to spot it, and how to subdue it. This process is related to some of the forms of digital resistance we've been exploring, from online racial checking in response to microaggressions, to double-sided consciousness and memes that expose the way that racism operates.

This innovative online-based resistance can lead to a particular type of consciousness-focused activism, which doesn't fit neatly into existing models of activism, organizing, or participatory politics. Online activists make posts meant to stimulate growth in racial consciousness and the ability to recognize and critique racist structures, attitudes, and behaviors. They see individual-level racial attitudes and frameworks as sites of activism, and part of their work includes breaking down the racial ideologies that further racial inequality, racial bias, and White supremacy. Their goal is to train their peers to use a critical racial lens when interpreting dominant social norms, public and political events, and interactions on campus. In this way, they seem to be following the transformative model of community organizing, which emphasizes building critical consciousness so that community members can recognize and challenge the dominant ideological frameworks that support unjust systems.[31] Triumph, for these students, does not need to be tied to a formal policy change, and instead can derive from the way their peers demonstrate or articulate growth in their understanding of the social meaning of race.

For these young activists, online racial checking is a tool, online counterspaces are the sites of resistance, participatory politics is the method, double-sided consciousness is an outcome, and organizing is the ultimate goal as young people of color invest in the consciousness development of their peers and networks.

8

RACISM IS TRENDING

I am not free while any woman is unfree, even when her shackles are very different from my own. And I am not free as long as one person of Color remains chained. Nor is any one of you.

—AUDRE LORDE, "The Uses of Anger," 1997

The Empire is a disease that thrives in darkness. It is never more alive than when we sleep.

—MAARVA ANDOR, *Andor*, 2022

THE VALOIS CAFETERIA is a well-known restaurant in Chicago's Hyde Park neighborhood, and is about a five-minute drive from President Obama's house. Legend holds that President Obama used to frequent the diner, and his old order, an egg-white omelet, is now advertised as the Obama Special. Imagine, just sixty years before he took office, Barack Obama would have been turned away from most restaurants in the United States, and hundreds of Black people were beaten and arrested for sitting down in cafeterias or diners like Valois. But from 2008 to 2016, Obama was arguably the most powerful person on the planet. How inspiring to see so much progress in a relatively short time. For many people, Obama's rise was strong evidence of the decline of racism in America and the beginning of the post-racial era.

Of course, people who think seriously about race and racism never believed in the myth of a postracial era. Racism has been masked and invisible, but not absent. In contrast with the notion that Obama's tenure as president is evidence of the end of racism, he has in reality been more of a lightning rod for racism, attracting an alarming amount of racial animosity. He was a symbol of hope and change, but not everyone wanted change.

Much has been written about the racist backlash to Obama's presidency. Scholar Carol Anderson uses the term *White rage* to refer to conservative policy pushbacks in response to perceived Black advancement, including President Obama's tenure in the White House.[1] Some conservatives used Obama's election as proof that racism was dead and argued that the Voting Rights Act of 1965, which sought to outlaw sneaky ways of keeping Blacks from voting, no longer needed to be enforced by the courts.[2] This backlash also found its way online. Not only did the number of online hate sites increase by 755 percent between 2008 (the year Obama was elected for his first term) and 2009, but online posts on hate sites after Obama's election also tended to be more violent than they were before he was elected.[3] Journalist Anthony Van Jones attributed Trump's election in 2016 to what he calls "Whitelash," a racist pendulum swing away from the symbol of change that Obama represented to a reclaiming of White domination: "Make America Great Again."[4] In 1992 Tupac sang (or rapped) the words, "Although it seems heaven-sent / we ain't ready to see a Black President." Given the backlash we've witnessed against Obama, it seems Tupac, ever the prophet, was right: sixteen years later, America still wasn't ready for a Black President.

INVISIBILITY LOST

In the days leading up to the 2016 presidential election, President Barack Obama said that a victory for Trump would be a direct affront to his legacy.[5] This is consistent with the way many Democrats and left-leaning Independents responded to Donald Trump's 2016 election as more than just a political loss; it was perceived to be a collective tragedy. One potential explanation for this is that Trump's racial rhetoric and conservative policies caused him to be perceived as the embodiment of old-fashioned racism: an ugly, open, and unapologetic racial ideology. As Trump was finding success on the campaign trail, David Duke, former leader of the Ku Klux Klan, said, "The fact that Donald Trump is doing so well, it proves that I'm winning."[6] Many liberals or antiracists would agree.

But in a society that explicitly values racial equality and where it is illegal to formally discriminate based on race, racism is most potent when it hides in the shadow of institutions and structures that use oft-hidden advantages, purportedly colorblind laws and policies, and surface-level egalitarian attitudes to create racial inequities. When racism is invisible, it is difficult for researchers, educators, activists, and citizens to identify its causes and effects, and it is even more difficult to demonstrate these causes and effects to White folks in a way that is both convincing and unlikely to put them on the defensive.

Open racism is repugnant for most Americans. The unspoken agreement to maintain a certain decorum—even while enacting violent, even murderous policies—is a crucial part of American politics. Trump's victory—perceived by many to be a win for White supremacy—was short lived for this reason: it came at the cost of racism's invisibility. Throughout his tenure in the White House, Trump experienced the consequences of being perceived as a clear threat to our country's values, and people from diverse viewpoints have come together in a sustained, anti-Trump movement throughout the country. He was impeached twice, and never had an approval rating over 50 percent.[7] With over 80 million votes, more people voted for Biden in 2020 than any other presidential candidate in history.[8] Biden wasn't a wildly popular candidate, and I think it's fair to attribute this turnout, at least in part, to anti-Trumpism: people voting more to get Trump out than to get Biden in.

And in 2021, after Trump's failure to be re-elected, he was banned from all major social media platforms, and the one platform that would accept him was banned from the major app stores.[9] Despite this backlash, and a myriad of ethical concerns with Trump's behaviors during his presidency, conservatives are torn regarding what role Trump and Trump-style politics will play in the Republican Party's future. Some are convinced conservatives need Trump's base to be successful, while others want to move away from Trump to either reclaim the party's historical values or reimagine the party's priorities. As the Republican Party makes these decisions, some conservatives in Congress have gained attention by rebelling against the establishment, Trump-style, by breaking the rules and refusing to wear a mask or disclose whether they had been vaccinated in the midst of the COVID-19 pandemic.[10] These acts of rebellion can make it seem that they are as independent from the political machine as Trump was believed to be. The backlash against Trump, including from social media companies, may not be permanent. In fact, as billionaire Elon Musk made a bid to purchase the Twitter platform in 2022, he announced plans to reinstate Trump's account, a decision allegedly tied to his beliefs around free speech.[11] Soon after Musk's acquisition, Trump's Twitter account was reinstated.

In 2021, Kamala Harris claimed that America was not racist, something that disappointed many of her followers who supported her as the first Black and Indian woman to be elected as vice president.[12] But to imagine why Harris might have made this claim (we can imagine without excusing), consider the uproar if the VP had suggested the opposite during a time when critical race theory was being banned in select places across the country. Most Americans don't believe racism is prevalent in American society and don't want to think about it. They dismiss incidents of racism as uncommon and do not want to hear that these incidents are evidence of a deeper problem.

Throughout this book I've argued that the unmasking of racism, largely in online spaces, provides new opportunities for resistance. It's easier to fight (and defeat) an enemy you can see. It's easier to convince allies to join the fight when the enemy is in clear view. Technology has provided the means to make racism more visible and more accessible to people who don't (knowingly) encounter it in their everyday lives. For most people, racism is relegated to history, one of the old gods who has lost its vitality. But now, it's becoming clearer to the masses that racism is as formidable as ever. Racism is trending. The longer it trends, the more people will see it unmasked.

Online discussions of racism are more common and becoming normalized. More attention is being brought to the invisible ways racism operates. As racism trends, as videos of racism go viral, as people on social media challenge racism—from influencers with millions of followers, to regular people posting for their personal networks—these acts expose the mechanisms of racism and prove the continuing significance of race. Racism trending is a sea change for a society that has been characterized by color-blindness in the twenty-first century.

The students I talked with told us about the increase in discussions of race on campus, something that is helping to change the campus racial environment, increase racial awareness, and challenge various forms of racism. But this book is not just about colleges and universities. How can we determine whether these trends are taking place on social media more broadly, and not just on campus?

To measure how much racism has been trending online, I conducted searches for race-related keywords on Twitter for over ten years, from 2011 to mid-2021. In this chapter, the last chapter, I will be discussing and presenting charts that visualize these trends, demonstrating just how common some race-related language has become in online spaces. My goal with these charts is not to inundate you with data, especially as the book is winding down. But it is important that you see that just as the conversation is shifting on campus, the conversation is shifting in broader online spheres as well. Take note of the shape of these charts, and see how they are all trending upwards— all except figure B, which looks at the total number of tweets during this time period (I included this chart just to show that the trends we see are not simply due to increased usage of Twitter—Total Twitter use was actually decreasing as usage of many of the race-related terms we explore were increasing).

The first term I researched is simply *racism* (or *racist*), shown in figure C. Mentions of racism or racist(s) have been increasing each year, spiking in 2020 with the death of George Floyd, which, on the heels of the murders of Breonna Taylor and Ahmaud Arbery, sparked the sustained 2020 protests and uprisings against anti-Black police violence. Use of *racism/racist* really picks up in 2016. Out of curiosity, I found that the average daily Twitter mentions of the term *racist* increased from 7,356 in the six months preceding Trump's nomination as the Republican candidate in July of 2016, to 13,774 in the six months after.[13] I speculate this has something to do with Trump's more openly racist policies and rhetoric, but I don't want to give Trump all the credit. The Movement for Black Lives, together with BIPOC activists and social media users, were driving this increase well before Trump's rise.

To account for the term *racist* being used to refer to old-fashioned, in-your-face racism, I performed another search, this time for *institutional, structural, and systemic racism*. These terms are less likely to be thrown about casually or as an insult. They typically imply a more advanced understanding of how racism operates. Increased mentions of structural, institutional, or systemic racism can be seen in figure D. Because thinking about racism as an individual-level phenomenon is the dominant way of thinking about racism, referring to racism on a structural level challenges and advances how most people think about racism.

If you already are accustomed to thinking about racism as a structural phenomenon, you may not think this shift in language is a big deal. But how did you begin to think about racism as a structural problem? Do you remember when your paradigm for understanding racism shifted from being about evil racists with pointy white hats, to being about policies and organizations that on a surface level have nothing to do with race? Was it a class you took, a thoughtful friend you stayed up late talking to, or a person you follow on social media? In my interviews, I found that many students point to various online sources that complicated and deepened how they think about and discuss racism. More and more people are exposed to this informal, online knowledge production and dissemination that has powerful consequences for our understandings of racism in the digital age.

There are other terms that are increasing in usage that are also designed to expose and teach about the ways racism works. Some of these have become so common they can be thought of as buzzwords. As with all buzzwords, they run the risk of being overused to the point their meanings can be watered down or become meaningless. An unfortunate example of this is the term *woke,* which once was a smart way of conceptualizing racial, social, and political consciousness, but today is more often used pejoratively by conservative voices criticizing the "woke crowd," those they perceive to be progressive on issues around race, gender, or orientation. But on the way to overuse lies a sweet spot, when the language is buzzing just enough to indicate a change in the base level of racial consciousness, as more people develop a vocabulary that makes it easier for them to understand and articulate the ways that racism works.

One example is the term *microaggression,* seen in figure E. In a 2020 paper, my colleagues and I did an in-depth qualitative analysis of how the term *microaggression* was being used on Twitter. One-third of the usages were anti-microaggression—people who thought microaggressions were not real or the term overused.[14] This means that the term is not just being used amongst like-minded, antiracist folks. This widespread adoption of the term, even among the opposition, means that the world is becoming more aware of, and talking about, the types of subtle experiences that support the continuation of racial oppression but typically go unnoticed or unchallenged.

Another buzzword is *White privilege,* which refers to unearned benefits Whites receive based on their Whiteness. Privilege, including not

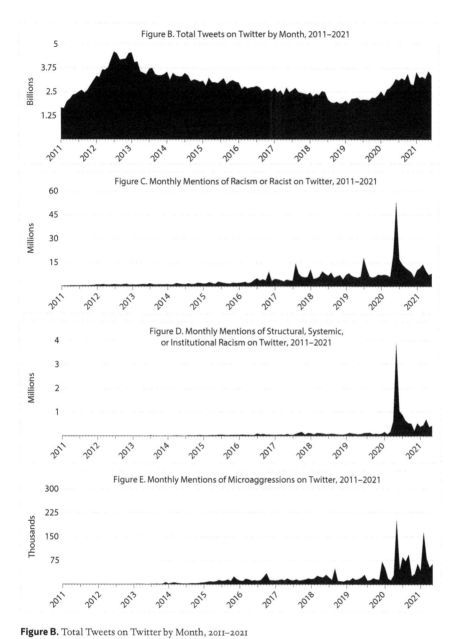

Figure B. Total Tweets on Twitter by Month, 2011–2021

Figure C. Monthly Mentions of Racism or Racist on Twitter, 2011–2021

Figure D. Monthly Mentions of Structural, Systemic, or Institutional Racism on Twitter, 2011–2021

Figure E. Monthly Mentions of Microaggressions on Twitter, 2011–2021

Figure B. Total Tweets on Twitter by Month, 2011–2021
Figure C. Monthly Mentions of Racism or Racist on Twitter, 2011–2021
Figure D. Monthly Mentions of Structural, Systemic, or Institutional Racism on Twitter, 2011–2021
Figure E. Monthly Mentions of Microaggressions on Twitter, 2011–2021

just White privilege, but also male, straight, cis, class, ability, citizenship, and religious privilege, can be difficult concepts to teach.[15] And it is possible (and common) to intellectually understand privilege as a concept but have a hard time grasping one's own privileges. Many of my students, and my interviewees, have discussed being exposed in online spaces to ways of thinking about privileges they didn't realize they had. The increased discussion of privilege (seen in figure F) can have an impact on people's ability to understand the way racism influences their lived experiences.

Another term that has increased in usage is *cultural appropriation,* which describes the assimilation of cultural symbols, practices, or resources of a nondominant group, by a dominant group.[16] It is this power dynamic that separates cultural appropriation from cultural exchange. Those who are having their culture appropriated have traditionally lacked the power to speak out against it, while those who engage in cultural appropriation often have the power to benefit or profit off it without consequence. But through social media, challenges and critiques to various forms of cultural appropriation have become more and more common. While the overall mentions of cultural appropriation increase over time, there are more visible single-instant spikes in usage for this term, seen in figure G, which may be explained by instances when celebrities engaged in very public forms of cultural appropriation.

The most convincing explanations for how online spaces came to be characterized by increased attention to issues of racial inequities start with Black women, a point Feminista Jones and Catherine Knight Steele make strongly in their books.[17] And for the Black women who have been driving Twitter culture, as it relates to critical challenges of racism, sexism, and homophobia, racism has never been just about race. For example, scholar Moya Bailey coined the term "misogynoir" on Twitter to refer to the specific brand of racism that is unique to Black women.[18] Bailey's book explores how Black women resist misogynoir in online spaces. It is clear that one of the defining characteristics of digital resistance is its intersectional concern with not only racism but also other forms of domination.

Perhaps the most powerful example of a buzzword that points toward a more critical way of analyzing racism is *intersectionality,* a term created by Kimberlé Crenshaw to explain the ways racism intersects with other forms of marginalization, including gender, sexual orientation, class, ability, or religion, to create distinctive forms of oppression.[19] Black feminist thought has long recognized the inescapable connection between racism and other forms of oppression. For example, in 1977 the Combahee River Collective, a radical Black feminist collective, wrote, "If Black women were free, it would mean that everyone else would have to be free since our freedom would necessitate the destruction of all the systems of oppression."[20] The quote from Audre Lorde at the beginning of this chapter speaks directly to this mandate that the freedom struggle be inclusive, not segmented; our fates are tied.

Figure H shows increases in mentions of the term *intersectional* or *intersectionality* over time. As racism is discussed more broadly, it is increasingly done with an intersectional lens, ensuring that we are not fighting racism in a way that only benefits straight Black men. Figures I and J show similar patterns for searches for *the patriarchy,* and *heteropatriarchy.* While not all feminist thought includes the struggles of Black women and women of Color—in fact, Black women on Twitter have been instrumental in identifying "White feminists," or those who claim to support all women but are primarily concerned with issues unique to White women, ignoring injustices particular to women of Color—still, this increase in mentions of feminism is indicative of the normalization of this critical framework challenging the power structures that privilege men over women.

Civil rights movement leaders who are taught in schools are generally straight, Black men: Martin Luther King Jr. and Malcolm X being the most prominent (Rosa Parks is taught but is not often framed as a leader). But as I briefly mention in chapter 6, Ella Baker, a

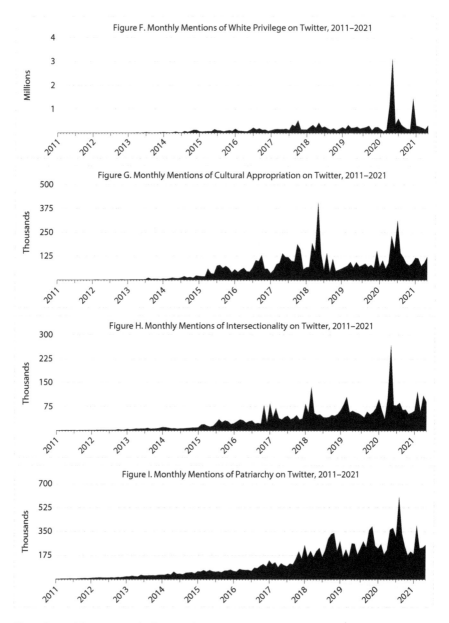

Figure F. Monthly Mentions of White Privilege on Twitter, 2011–2021

Figure G. Monthly Mentions of Cultural Appropriation on Twitter, 2011–2021

Figure H. Monthly Mentions of Intersectionality on Twitter, 2011–2021

Figure I. Monthly Mentions of Patriarchy on Twitter, 2011–2021

Figure F. Monthly Mentions of White Privilege on Twitter, 2011–2021
Figure G. Monthly Mentions of Cultural Appropriation on Twitter, 2011–2021
Figure H. Monthly Mentions of Intersectionality on Twitter, 2011–2021
Figure I. Monthly Mentions of Patriarchy on Twitter, 2011–2021

Black woman, and Bayard Rustin, a gay Black man, were two of the most important figures of the civil rights movement, with Rustin being one of the architects of the nonviolent protest strategy and writing some of King's best known speeches, and Baker doing the organizing work for decades that built the scaffolding needed to make the movement successful.[21] The BLM network, on the other hand, "centers those who have been marginalized within Black liberation movements."[22] Women and queer folks no longer need to take a back seat to straight men in leadership, and the analysis of and strategizing against anti-Black police violence is expected to be intersectional from the jump.

Some mentions of race-related terms seem to coincide with Trump's time in office between 2016 and 2020. For example, mentions of *dog whistle* (figure K) do not increase gradually between 2011 and 2016, like many of the other search terms, but instead seem to jump in 2016, with 12,856 tweets in June, followed by 29,658 in July and 37,063 in August (the months after Trump won the Republican nomination for president). Just as a dog whistle sounds at a frequency that is audible to dogs but not humans, dog-whistle politics refers to politicians using language designed to communicate with a specific group of constituents without alienating other voters.[23] More people pointing to dog-whistle politics can be the result of general increases in racial consciousness indicated by other charts, but likely also comes in response to Trump's campaign and administration: Trump's racist dog whistle operates at a different frequency than those of most Republicans. Whereas the traditional conservative dog whistle is designed to avoid detection, masking racist policies behind fiscal or value concerns, Trump seemed to choose to use a "regular" whistle, audible to humans and so loud it can give people a headache. His policies are harder to be explained as being driven by fiscal conservatism, instead of racism.

Similarly, in figure L we see that mentions of *vote(r) suppression* began to increase around 2016, with the monthly average mentions increasing

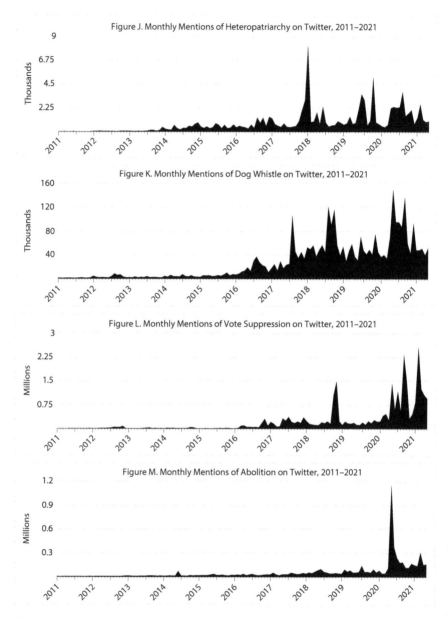

Figure J. Monthly Mentions of Heteropatriarchy on Twitter, 2011–2021
Figure K. Monthly Mentions of Dog Whistle on Twitter, 2011–2021
Figure L. Monthly Mentions of Vote Suppression on Twitter, 2011–2021
Figure M. Monthly Mentions of Abolition on Twitter, 2011–2021

from 10,166 in 2015, to 87,083 in 2016, and up to 1.2 million monthly mentions in 2021. The 2021 increase came at the same time conservatives around the country enacted voter repression policies to reduce Black and Latinx voter turnout, which some believe is the only way to keep the Republican Party viable.[24] I first learned about this recent voter suppression push through an Instagram post, which was itself a screenshot of a Twitter post. These posts prompted me to do further research. Following activists on social media can make it easier to keep track of justice-related issues, especially those not being reported by mainstream media. The chart in figure L represents on a large scale what many of us have been experiencing on our personal social media feeds.

The most striking trends, in my view, are seen when examining mentions of more radical terms on social media, things such as *prison abolition* or *reparations*. Books like Angela Davis's *Are Prisons Obsolete?*, Michelle Alexander's *The New Jim Crow*, and Reuben Miller's *Halfway Home*, have brought into public consciousness the realities of the criminal justice system, its racist nature, and its impact on the lives of incarcerated and formerly incarcerated folks.[25] There are many schools of reform, with some calling for shorter sentences and more programs to prevent prison re-entry, and others demanding the release of those incarcerated for drug offenses, especially now that weed can be bought and sold legally. White people, including the billionaire governor of Illinois, J. B. Pritzker, are getting rich off weed, while far too many Black people are still locked up because of it.

Prison abolitionists question the very logic of carceral systems—from prisons to courts to the police.[26] This is a political ideology that is often seen as being far-left, and can make many left-central–leaning proponents of prison reform uncomfortable. Prison abolitionists are intentional with the language they use to describe the prison-industrial complex in order to challenge dominant ways of viewing punishment, and they encourage us to imagine a world where jails are neither seen as justice, nor needed.

I believe the gradual increase in mentions of abolition we see in figure M, with increases between 2013 and 2021, is connected to the Movement for Black Lives. Figure M shows an explosion in the use of *abolition* in 2020, with tweets mentioning abolition jumping from 42,066 in April, to 109,219 in May, to 1,154,777 in June. This explosion takes place at the height of the uprisings protesting anti-Black violence: on June 6, 2020, 500,000 people took to the streets at over 500 sites in the United States.[27] Many leaders and activists associated with the uprisings against anti-Black police violence utilize an abolitionist framework, suggesting that police reform efforts have failed and demanding a reimagining of what public safety looks like in a world without police. Even as in-person protests slowed down towards the end of 2020 and beginning of 2021, the mentions of abolition on Twitter increased.

Similarly, figure N shows tweets with the terms defund or *abolish* police. Big cities often spend a disproportionate amount of their budgets on police departments, including a quarter of the city's budget going towards police in Los Angeles and Baltimore, a third in Houston and Chicago, and over 40 percent in Oakland and Milwaukee.[28] The call to defund the police, reinvesting the money in programs that would reduce the social problems that often precede involvement with the criminal justice system, including poverty, food scarcity, mental illness, or joblessness, originated as an abolitionist demand during the Ferguson protests.[29] An example of a program associated with this demand would be an alternate means of emergency response, like sending trained mental health professionals to help someone having a mental health crisis instead of the police. A social worker trained in anticarceral methods would respond differently to this crisis than a police officer would, seeking to get the person the help they need, rather than handcuffing or beating them or sending them to jail when they really are in need of assistance.[30] Groups like #8ToAbolition continue to organize for "measures that reduce the scale, scope, power, authority, and legitimacy of criminalizing institutions," with the ultimate goal being the elimination of the police institution.[31] Other supporters of defunding the police may want to see a reduction and redistribution of police budgets coupled with police reforms, but do not go as far as demanding abolition.

We find similar patterns when we search on Twitter for the term *abolish ICE,* as seen in figure O. ICE refers to the Immigration, Custody, and Enforcement Agency, a division of the Department of Homeland Security. The #AbolishICE hashtag came into prominence in 2018, when the separations of migrant families at the southern border of the US, and children in cages, made national headlines.[32] Left-leaning activists responded to these images not by simply suggesting an ICE policy change, but by demanding the entire organization be abolished, given its history of abuse and terror that long preceded the Trump administration.[33] This hashtag comes from an anticarceral, abolitionist approach to thinking about immigration reform that critiques not just Trump but also Obama, Clinton, and Bush before him. I believe it was easier for the media and the public to be outraged at Trump's family separation policy, because of (1) widely spread and distasteful images of children in cages, and (2) Trump's history of racist remarks towards the Latinx community, which made the connection between his policies and bias more obvious. But the calls for abolishing ICE recognize that the origins of the mistreatment of people of Color at the border do not begin with Trump, and is something both liberals and conservatives have been guilty of in the name of protecting the border (language that we discuss as part of the Latino threat narrative in chapter 2).

Another radical form of antiracism can be seen in figure P, in the term *reparations.* Reparations are defined by the Movement for Black Lives as the "act or process of making amends for a wrong," or restoring a group injured by human rights violations based on race (or another form of identity).[34] Some examples of reparations include the billions of dollars Germany has paid Israel in reparations for the Holocaust, and money the United States has paid to Japanese Americans who were placed in internment camps during the Second World War.[35] In America, this discussion typically centers around reparations for slavery. Freedwomen and freedmen were meant to be given reparations in the form of 40 acres and a mule (which would have allowed them to make an agrarian living), but this was rarely done. Instead, many freedwomen and freedmen were forced into sharecropping, a form of labor and economic exploitation.

In the Unites States, reparations for slavery can seem like a pipe dream. Giving reparations implies that there is some contemporary culpability for historical injustices, something that flies in the face of rugged American individualism, which includes the notion that we are each responsible for our own choices, and a minimization of the effects of historical wrongs on our current circumstances. As an example of this, scholar Robert Patterson notes that one of the biggest jobs for activists fighting for reparations is to explicitly connect government actions during slavery, Reconstruction, and Jim Crow to the current social and economic realities in the Black community—something that is difficult because conservatives and centrists prefer to ignore the history of racial oppression and have a hard time admitting its contemporary relevance.[36]

But figure P shows that online discussions of reparations have been increasing, including an initial jump in 2014, followed by five years of gradual increases, and an explosion in 2019 that has remained relatively constant through 2021. At the same time that Twitter discourse of reparations has been increasing, we have seen at least eleven mayors across the United States engage in what we might call proof-of-concept reparations efforts, using local policies and funds to engage in experimental efforts at giving reparations to Black people living in communities that have benefitted from slavery and have engaged in discrimination against Black residents.[37]

We can't causally point to Twitter as the sole reason reparations are taking hold. Activists had been demanding reparations long before

Twitter commanded the interests of politicians and the mainstream media. And the local activists and policy-makers that pushed for reparations in Illinois, North Carolina, California, or Colorado may not have been on Twitter. But the more a topic is discussed on social media, the more it is normalized, the more it makes the news—the more policy changes such as reparations become palatable for voters and policy makers. And because politicians pay attention to things that trend on social media, they may seek to adopt policies that ingratiate themselves with the Twitter users behind the trending issues.

I've always believed that giving reparations to Black people was just. Not just reparations for slavery, but also for the centuries of laws and policies that limited our life chances and built the structures that make it so hard to escape poverty today, from Jim Crow laws to redlining and the willful creation of the ghetto; and for the discriminatory policies that kept Black folks from benefitting from government programs that helped White Americans build wealth, like the G.I. Bill of Rights (1944) that subsidized home ownership for White veterans but excluded Black veterans.[38] But I never imagined reparations becoming real. If I have a hard time convincing my ostensibly nonracist White friends that Black people deserve reparations, how could we possibly convince the world or people in power? Social media discussions have been contributing to this "imagination" work, helping people envision a different world where Black people and people of Color are valued, and society is accountable for the past wrongs that shape our contemporary realities. This increase in discourse doesn't mean reparations will become a reality for Black people on a national level anytime soon. Yet the increased discourse—as well as the proofs of concept and activist victories around the country—let us know that reparations are no longer laughable, the way Jerry describes the reaction to his talking about reparations in chapter 3.

In June of 2021, President Joe Biden made Juneteenth a national holiday. Juneteenth is a day that celebrates the Union army reaching Texas several months after Confederate General Robert E. Lee surrendered and the end of the Civil War, and declaring that all slaves were free. Frederick Douglass's famous speech, "The Meaning of the Fourth of July for the Negro," explored the contradiction of Independence Day, as on July 4, 1776, only White folks were free, and Black folks were still enslaved.[39] The celebration of Juneteenth acknowledges the contradiction of Independence Day and highlights the way the fight for freedom extends beyond the Revolutionary War, or even the Emancipation Proclamation.

Growing up, each year my extended family celebrated Kwanzaa, a holiday that recognizes Black history and African heritage. Despite this commitment to celebrating Black heritage and culture, I did not hear of Juneteenth until I was an adult. In figure Q we can see bumps in conversation about Juneteenth each year that increase in size across time, from 63,107 in June of 2016, to 204,129 in June of 2017, to over 6.3 million in June of 2020 (in the midst of the uprisings), and 4.1 million in June of 2021, when Biden made the holiday official. Twitter can put things like reparations, or Juneteenth, on the radar, which can influence the types of policies politicians and folks in power pursue. It's not only Twitter, of course. Activists like Opal Lee have organized in-person demonstrations and events to spread the word about the holiday for years.[40] But online rhetoric and discourse shapes the discussions that are had from traditional media to Capitol Hill. In the case of Juneteenth, face-to-face activists and online users both contributed to speaking and marching this holiday into existence.

At the same time justice-oriented topics trend on social media, these platforms can also be used to spread hate. Figures R and S show that usage of #BlackLivesMatter (BLM) and #AllLivesMatter (ALM) have both been increasing in popularity over the past decade (although BLM has been used nine times more than ALM, or around 92 million more times than ALM over this time period). Seeing people tweet or talk about ALM can be disheartening, and it is difficult to discern whether

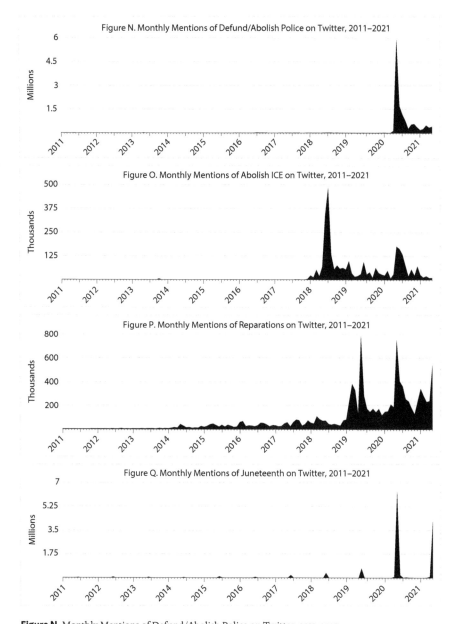

Figure N. Monthly Mentions of Defund/Abolish Police on Twitter, 2011–2021
Figure O. Monthly Mentions of Abolish ICE on Twitter, 2011–2021
Figure P. Monthly Mentions of Reparations on Twitter, 2011–2021
Figure Q. Monthly Mentions of Juneteenth on Twitter, 2011–2021

those behind these messages are open racists using ALM as cover or confused citizens who need to be schooled on what BLM is.

But, when *either* All Lives Matter or Black Lives Matter trend, *both* trends challenge the dominant narrative in different ways. The dominant mode of understanding racism suggests that racism is an individual-level problem that largely died off after the civil rights movement. BLM tweets challenge this narrative by highlighting the depth of racism in US structures and institutions (not just the police). At the same time, the ugliness and ignorance associated with ALM proves that racism is not dead. As the naming of contemporary expressions of racism increases, so does the collective capacity to recognize racism in its myriad forms, even the forms of racism largely hidden from mainstream eyes. Similarly, but perhaps counterintuitively, online racism also alerts online users to the widespread continuation of old-school style racism—a phenomena discussed at length in chapter 4—which can put them on alert for racism from unexpected sources.

Sometimes these long-term shifts in online racial discourse are bolstered or jump-started by more short-term, viral responses to racism or race-related issues. For instance, the hashtags #StopAsianHate and #StopAAPIHate were used over 4.7 million times between their inception in March 2020 and June 2021. These hashtags brought attention to disturbing increases in anti-Asian hate crimes between 2019 and 2022. Some of my Asian friends and colleagues shared the ways they and their family and friends were on the receiving end of verbal attacks, from being called "corona," to being treated rudely in ways that were not typical for them before the pandemic. As I briefly mention in the last chapter, this type of anti-Asian bias appeared to be sanctioned at the highest levels: because the COVID-19 virus first hit the news as an outbreak in Wuhan, China, then-President Trump insisted on referring to the virus as the "Chinese virus," going as far as to cross out "corona" and write in "Chinese" on a speech that had been prepared for him.[41]

Things seemed to come to a head in early 2021 when a number of attacks on the AAPI community made the news, from the San Francisco Bay Area to New York City, followed by a mass shooting at an Atlanta massage parlor. While some mainstream news outlets found nonracial rationales for the murders (including religion, mental health, and stress), social media users challenged these dominant narratives and insisted that anti-Asian racism be recognized and named.

I also remember that for a time, most of the media images and videos of anti-Asian violence seemed to be perpetrated by Black folks.[42] If mainstream media were to be believed, it appeared the increase in anti-Asian violence was being driven by the Black community. Black activists around the country spoke out against this violence right away, and many Black people began to ask hard questions about what might be going on in the community. But research has shown that around 75 percent of anti-Asian hate crimes are perpetuated by White people.[43] Despite this media misrepresentation, the public examples of anti-Asian violence at the hands of Black people led to activists from both communities working together to rally against these acts of violence and come up with solutions.[44] Online discussions put anti-Asian violence into historical context, detailing the history and contemporary realities of Asian-Black relations, using White supremacy as a frame through which to understand both anti-Asian attitudes and actions in the Black community, as well as anti-Black attitudes and actions in the Asian community, and challenge the idea that the best solution would be increasing police patrols (which could be seen as increasing anti-Black state violence in order to combat anti-Asian violence).

One example of online users thinking critically about these events and anti-Asian attitudes more broadly is the increased discussion online of the term *model minority myth,* seen in figure T. The model minority myth refers to the way Asian Americans are seen as the model minority: well-behaved, high-achieving, and self-sufficient, often in contrast to stereotypes of other minoritized groups, like Black, Latinx, and

Indigenous folks as being criminal, low-achieving, and dependent.[45] While in some situations this myth may seem to benefit Asians, a critical analysis of the myth finds that it perpetuates stereotypes of Asians as foreigners, which in turn normalizes the type of anti-Asian sentiment that was often ignored prior to 2020. Usage of the term *model minority myth* jumped from 5,033 in April of 2020 to 39,872 in May of 2020 and 41,157 in June of 2020. Mentions died down during the second half of 2020, but peaked in March of 2021, the month of the anti-Asian mass shooting in Atlanta, with 57,316 mentions. The model minority myth is an academic concept that has been increasingly used in online spaces to boost public understanding of the ways anti-Asian bias, a product of White supremacy, operates in the United States, from presumably benign stereotypes to horrific acts of violence.

In 2021, Palestine went viral. Social media was filled with videos depicting several Palestinian families being evicted from their homes in the Sheik Jarrah neighborhood in East Palestine, protests in response to these forced evictions, as well as Israeli aggression towards protestors and residents, including shooting tear gas, rubber bullets, and bullets into a mosque and beating worshippers.[46] These videos were, for many, an introduction to "settler" politics in Palestine. In one viral video, a Palestinian woman pleads with a settler named Jacob, saying, "you are stealing my house." His response was to say, "And if I don't steal it, someone else is gonna steal it," as if this absolved him of all guilt.[47] Other videos included explanations, putting into layman's terms the types of evictions that have been going on for decades in Palestine: as Israeli courts determine that Palestinians are staying on land that belonged to Israelis before 1948, now, more than seventy years later, Palestinians are being forced to leave family homes they have lived in for generations.

Figure U shows mentions of the word *Palestine* on Twitter between 2011 and 2021. There are small bumps in usage of the term over time, corresponding to moments when conflict in Palestine/Israel was making international headlines, including the 2014 Gaza War. But in May of 2021, we see a huge increase, with a little over 300,000 tweets in April of 2021, but over 18 million in May of 2021. This is both unique and significant, given the hesitance many folks feel about speaking up about injustices in Palestine. Historically, speech in support of Palestine has often been labelled antisemitic, and there are numerous examples of Black activists, artists, and scholars whose careers were harmed because they were accused of antisemitism as a result of their support of Palestinian freedom or their critiques of the Israeli occupation.[48]

For example, when scholar and former CNN correspondent Mark Lamont Hill gave a speech in support of Palestine, there was a media firestorm and he not only lost his job at CNN, but there was talk about him potentially losing his position as a tenured professor at Temple University.[49] And legendary activist Angela Davis had an award from the Birmingham Civil Rights Institute (temporarily) revoked as a result of criticism of a boycott of Israel she supported.[50] In his book, *Vibrate Higher,* rapper Talib Kweli talks about his shows in Germany being cancelled after he attended an event supporting Palestine.[51] Despite cogent arguments that insist criticizing Israeli policy is not antisemitic, and that explain why we cannot conflate "a diverse religion with the politics and policies of a single country," speaking up for Palestine can have consequences that may scare away some supporters.[52]

In 2021, activists chose to call out this fear. In addition to heart-wrenching videos, images, and interviews, my personal social media feeds were filled with messages from activists encouraging others to repost the videos that weren't making the news. Many of these posts suggested that being silent about what was happening in Palestine was being complicit in the oppression of Palestinian peoples. These informal prompts or pushes from activists may have helped accelerate the rate at which news from Palestine was being reposted.

Perhaps these activists were too effective. Activists and people reposting activist and primary source content began to report that

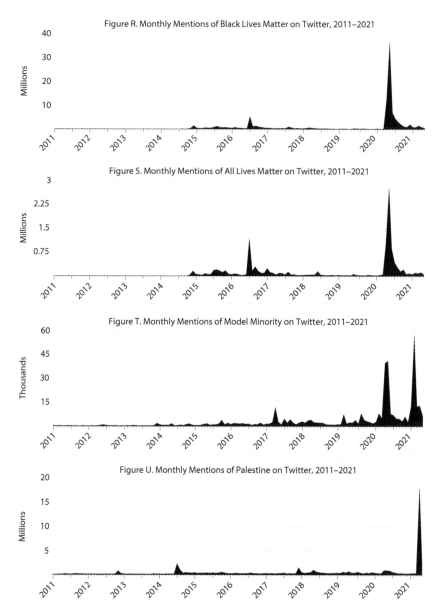

Figure R. Monthly Mentions of Black Lives Matter on Twitter, 2011–2021
Figure S. Monthly Mentions of All Lives Matter on Twitter, 2011–2021
Figure T. Monthly Mentions of Model Minority on Twitter, 2011–2021
Figure U. Monthly Mentions of Palestine on Twitter, 2011–2021

their posts were being blocked and removed by Facebook.[53] One of my friends, who had posted more than twenty videos from on the ground in Palestine to her social media stories, posted a screenshot of a message from Instagram saying that she would no longer be allowed to post on the subject. It seemed that as Palestine was going viral, social media companies were limiting the spread of images, videos, and commentary regarding what was going on in Palestine. Of course, these companies denied the banning was intentional and blamed malfunctioning algorithms.[54] Whether or not you choose to believe their explanations, this is an important reminder that though social media can be used to bypass traditional media gatekeepers in promoting issues of racial justice the powers that be do not want the public to hear about, social media companies have also become gatekeepers themselves.

FINISHING THE PUZZLE

The longest I've ever gone without seeing another person of Color was the week I spent in Maine working to finish this book. I was in a cabin that was twenty minutes away from the nearest grocery store and coffee shop, way out in the sticks. It was a wonderful week in January, and besides arriving at the start of a snowstorm,[55] sliding backwards down the snowy hill my AirBnb sat atop,[56] and being worried that I would run out of logs to burn,[57] it was a wonderful trip. Every day I wrote by the fireplace and took breaks in the outdoor hot tub, surrounded by a snowy forest with a frozen lake visible in the distance. The second day a truck came by and plowed away most of the snow in the driveway and on the hill, and I (bravely, I thought) ventured into town to get some groceries and a latte.

As I drove, I noticed a Confederate flag hanging near another flag that I didn't recognize. Because I associate the Confederate flag with racism, I wondered whether this other unknown flag was also advertising White supremacy. My AirBNB had a sliding deck door that didn't lock, which already felt uncomfortable for this native Chicagoan (I've always been amazed at my suburban or rural friends who don't feel the need to lock their doors). But the combination of a broken lock and (potential) White supremacists in the area was too much for me, and I decided to do a Google search for White supremacist groups in Maine. To my relief, there weren't many, but I did find myself on the website for a White supremacist organization with a local chapter. The group believed that because the United States was too diverse for Whites to maintain power via democracy, it was time to turn to fascism and take power by force.

I was initially surprised by this logic, because I had always associated White supremacist groups with hyperpatriotism and religious-level dedication to American freedoms. Fascism doesn't fit with a religious-level devotion to democracy. Then I noticed how this White supremacist logic was remarkably consistent with the way conservatives have been enacting voting restrictions to keep Black voters from casting their ballots.[58] Many conservatives believe that current demographic trends—with White people quickly becoming the numerical minority—will make it difficult for them to stay in power, and are therefore seeking to change the rules so that only people who think like them have access to the voting booth. But while Republicans are pretending to do this to prevent voter fraud, White supremacists are open with their desire to prevent Black people and people of Color from voting. As the logic of White supremacy becomes more visible, its connection to conservative efforts become clearer. This connection, such as what was seen with the January 2021 Capital riots, will weaken conservative efforts to maintain racist policies and practices without resistance.

THE FINAL PIECE

As racism trends, hidden racist attitudes and mechanisms continue to be revealed. But activists and antiracists, empowered by online communication, are refusing to let racism go unchallenged. The hood has come off, and racism is showing (and being forced to show) its bared face—at scale—for the first time in over half a century. This brings us back to the intellectual puzzle I began with: what does racism online tell us about racism in the real world?

It reveals it, exposes it, and opens it up to the types of questions racists are able to avoid when they adopt more covert methods. Online racism contradicts the dominant narrative and the postracial myth. It levels the public discourse playing field, circumventing the rules that have been put in place to limit how much we know and talk about racism. It also reminds people who thought racism was in our rearview mirror how much it still hurts today.

But there's another question, one I didn't know to ask when I began this project. How does online resistance fit in with the fight against racism in the real world? The rhetoric has changed, and the fight against racism looks much different now. The enemy is clearer and our tools more precise. More people are familiar with the language we need to use to describe racism, from the subtle microaggressions that characterize everyday interactions to the structural racism that can invisibly constrain our life choices and chances. And people are more willing to challenge racism, in part because social media has changed the racial power dynamics that have long silenced Black folks and folks of Color.

In an interview with filmmaker Spike Lee, comedian Eddie Murphy tells a story about a conversation he had in his office with boxing legend Mike Tyson and his then-promoter Don King.[59] When the conversation turned to White people, Eddie noted that the three of them began to whisper. Here were three of the most powerful Black public figures in the world, sitting in Eddie's private office—not around any White folks—but they were worried about speaking about race out loud. This is illustrative of the perceived danger of speaking out against White supremacy, even in a private setting, even exclusively among other Black folks, and even among individuals whose fame and financial success may insulate them from some forms of racism. For example, we might expect a police officer who stops Eddie Murphy to let him off with a warning, once he sees it is a celebrity behind the wheel. Comedian Dave Chappelle tells of this exact type of special treatment; a White officer let him go because of his celebrity status. The story ends tragically: the officer who stopped Chappelle shot a Black child the very next day.[60]

Today, Black public figures probably feel a bit different when discussing racism. Celebrities, including athletes, may feel less pressure to keep their mouths shut—or shut up and dribble—to make themselves attractive to corporate sponsors. Social media can provide them with a voice that is not controlled by corporations, which some athletes have even used to critique the corporations that pay them. Today, Black athletes speaking out against racism, or in support of the Movement for Black Lives, may receive more attention—and a bigger social media following—than those who are silent. Of course, there can still be consequences for resistance. Notably, football player Colin Kaepernick lost his job in the NFL after kneeling during the national anthem to protest the persistence of anti-Black police violence in the United States. But Kaepernick has not only been part of the change—the discussion around the unjust backlash against Kaepernick has made it easier for other athletes to speak out without experiencing the same consequences—but he has also benefitted in some ways from his protest, from receiving a Nike shoe with his likeness on it (something he was not on track to receive as an athlete), to being widely lauded for his activist work.

But social media is not just a tool for the rich and famous. Technology has bolstered resistance against racism by emboldening Black folks and folks of Color and amplifying our critiques of racist practices, people, and structures. Talking about racism at work can get you fired. But in addition to some folks of Color feeling more safe talking about racism online, when racist acts hit the internet, racists are the ones who need fear getting fired. This is a shift in power. The charts in this section reflect some of these promising changes in rhetoric. People are more comfortable talking about racism. This shift in language scares the opposition, which is why they are fighting so hard against theories about racism (like CRT) that name the mainstream institutions and policies that perpetuate racial oppression.

This is why interpersonal challenges to racism are so important. Those who are intentionally racist will say and do as much as they can get away with. And people who say or do racist things without realizing it won't stop until we point out the ways in which the things they say are problematic. Every racist act or statement that goes unchallenged makes the next racist act easier to get away with, because people continue to believe those acts of racism are normal, harmless, and permanent. Challenging racism, on the other hand, keeps racism from reproducing. It creates norms and environments that are hostile to racist acts, attitudes, and practices. The more spaces we make antiracist, the harder it will be for racism to get comfortable in our homes, schools, or places of work. Speaking out against racism makes it more difficult to be racist, intentionally or unintentionally. Of course, not everyone has the energy to speak out against racism, and that's okay. Maintaining your energy and mental health is itself a form of resistance.[61] But when people do speak up, it can be a powerful way of fighting racism.

At the same time, we can't overestimate the power of interpersonal challenges to racism. Interpersonal challenges should become a part of our everyday repertoires, but they can't be the entirety of our resistance. Our overarching goal remains changing the structures that limit the life chances of folks of Color. Some power structures have been the same for decades and appear to be immovable. But as social media shifts the public discourse, changes that once felt impossible can become realistic policy goals. We've seen several examples of ideas that long seemed outlandish to many go viral and then be implemented in the early 2020s, from defunding the police and redirecting money towards social workers, to actual reparations for slavery. So when activists call for the impossible, such as abolishing prisons, don't doubt them. Believe them, learn, and join the work!

We are seeing a fundamental shift in the way people understand racism in America, and it's still in progress. We've been taught to think that race is a permanent, fixed, immutable characteristic. But it's not. Race is dynamic, changing every day, and social media users are on the cutting edge of changing how we think about racial categories, racism, and injustice. As a race scholar, it's hard to keep up. By the time I identify a new phenomenon for study, get approval to run the study from the university research ethics board (IRB), gather and analyze the data, write up the results, submit the paper for publication in a scholarly journal, and then years later the paper comes out, the "new" phenomenon I set out to study is no longer new. On social media, however, the dynamism of race and racism can often be seen in real time.

While we've thought about social media as a tool that can empower activists, social media companies are far from perfect. TikTok users have reported their content being flagged as inappropriate when discussing Black Lives Matter, and as we discuss earlier in the chapter, Instagram users reported their accounts being limited when they were posting about Palestine in early 2021.[62] In 2021, Twitter implemented a bounty on coding bias in the app, encouraging hackers, researchers, and coders to find the ways their machine learning algorithms are inadvertently biased.[63] We have to continue to apply pressure to the social media giants, lest we exchange one set of institutional gatekeepers (traditional media) for another (social media).

Change is almost always scary, even when you expect a positive change. A world without racism would be dramatically different than the one we have now. We should expect the change to be resisted by both liberals and conservatives. We should expect calls for changes that have more bark than bite, policies and symbols that are meant to placate protesters without offering any legitimate solutions to racism. And sometimes, we should expect change to hurt. This is hard. Getting rid of racism sounds good to many, in theory. But fewer people are willing to stay committed if it costs them something. In the United States, we have been trained to think about things in individual terms—from rights to accomplishments to resources. We are all fighting to have enough for ourselves and our families or communities to survive. Even when we have excess, it may not feel like enough. There's no way around the reality that capitalism exerts control over what we think is possible (or desirable) in terms of racial justice. Activists, educators, and anyone who wishes to join the fight against racism are now faced with the task of directing these cultural changes and justice-oriented momentum towards the dismantling of the systems that enable racism, both overtly and covertly, even when such people are privileged enough to benefit from some of the existing power structures. There is power in the collective, and this process will be easier to accept when we consider the greater good to be as important (or more important) than private goods.

Other times we'll experience setbacks concurrent with apparent victories. For example, at the same time as we witnessed some semblance of accountability with Police Officer Derrick Chauvin being sentenced for the murder of George Floyd in June 2021, there were organized attempts at pushing back against racial advancement, from rolling back voting rights to the coordinated attack against critical race theory. Trump lost the election in 2020, but conservatives packed the Supreme Court. There is no way to know how far the Supreme Court intends to set back our clocks of justice. For many people, this is a scary time. But those interested in upholding White (male) supremacy are scared, too, which is why they have to resort to dirty tactics like voter suppression. As more people challenge their methods and policies, making it more difficult for them to stay in power legitimately, they will increasingly rely on these dirty tactics. This is not a reassuring thought, as it is not clear how far towards fascism they are willing to go to maintain control. But we should expect backlash to changes in racial power dynamics and challenges to the racial status quo, and be prepared to fight, from the polls to the courts to the streets.

Highly publicized and visible incidents of racism being punished may be used to further the myth that we are all on the same side, and that unmasked racists are rare: bad apples, not evidence of a broken and biased system. We must celebrate symbolic victories without being satisfied by them. As we continue to highlight the incongruities between American ideals and the realities of racism, we stretch the global imagination, introducing radical possibilities for reorganizing power and ridding our institutions and ideologies of racist cancers.

As racism continues to trend, we will have to contend with what it means for antiracism to be *trendy*. Corporations have an incentive to not be *seen* as racist. Whether or not antiracism is central to how an organization does business, it is now in its best interest to create antiracist programs, campaigns, and think carefully about the racist messages it may be inadvertently supporting in its ad campaigns. Some may not see this as being positive, as corporations appropriate terms and concepts from the resistance, or use antiracist terms and surface-level campaigns to mask their racist and oppressive policies. It is my belief that while we need to continue to critique these entities, we also should not miss the opportunity to use this sociopolitical moment to create change within capitalist organizations and institutions that do not have an incentive to fight racism outside of this external pressure.

Sports fans are notoriously loyal. Many remain faithful to their teams through thick and thin, even when their teams are losing and management and ownership seem committed to driving the franchise into the ground. For this reason, many loyal fans look down on fair-weather fans, or those who only join the fandom when their team is doing well or having a good year—when being a fan is cool. Those aren't true fans; they are just hopping on the bandwagon.

Because fighting racism is trending and seen as cool, we will undoubtedly continue to see more people hop on the resistance bandwagon. Those of us who were fighting racism before it became trendy may be tempted to treat bandwagon antiracists the way ride-or-die sports fans treat their fair-weather counterparts. But we shouldn't. Racism has been skillfully hidden in mainstream institutions and disguised in ways that can make it difficult even for well-meaning people to recognize it. It makes sense that most people can't see it—or couldn't, until its unmasking. But now that we have taken off the hood and racism is no longer secret, let's not shame people for taking so long to see the truth. Let's put them to work.

Acknowledgments

I RECEIVED A CONTRACT for this book in January 2020, a few months before the pandemic began. It was meant to be finished one year later, in January 2021. But I did not work on the book between March and November 2020. During that time I did my best to navigate life in quarantine, working on being the best father and coparent I could be, supporting my kids as they learned remotely, keeping the dishes clean, and making sure they got outside everyday after spending most of their time on Zoom with their classmates. The book was something that required more intellectual and emotional energy than I had during those first months of the pandemic. Even when my children were with their mother, and I had more time, the best I could manage was to keep up with collaborative projects, papers, teaching, and administrative duties.

When I reflect on this time, I am reminded of J. R. R. Tolkien's words in the preface to his *Lord of the Rings,* as he explained the long intervals between beloved installments, saying, "I had many duties that I did not neglect, and many other interests as a learner and teacher that often absorbed me." I am grateful to my editor, Naomi Schneider, and the University of California Press for allowing me flexibility in deadlines as I navigated fatherhood, academia, and finding the creative, intellectual, and emotional space needed to finish this project. This has been a challenging time for many of us, yet I know my family and I have had it easier than most.

I am grateful to my family, who provided the support and foundation that enabled me to embark on and finish a project like this. This book is dedicated to my children, Malachi, Karis, and JD. These three are my inspiration, my biggest supporters, and my most consistent time managers. My mother, Reba, who read my sister and I the *Chronicles of Narnia* as children and sparked a life-long love of reading, and did not know that reading and writing would one day become our professions. My father, Bob, who demonstrated an applied passion for justice and raised me to think beyond my own self-interest. My little sister, Reese Eschmann, author of *Etta Invincible* and the *Home for Meow* series, whom I look up to in so many ways—may the Black Inklings never die! Seeing your journey in publishing has been such an inspiration, and I'm so proud of you.

For the first four years after finishing my PhD, I was an assistant professor at Boston University, where I received much support and built lasting relation-ships with my amazing colleagues, both in my department and across the University. I was part of a junior-faculty reading group in the Sociology department that included Jessica Simes, Heba Gowayed, Saida Grundy, Joe Harris, Sanaz Mobasseri, Nicolette Manglos-Weber, and Nicholas Occhiuto. I presented early versions of chapters 6 and 8 to this group and benefitted greatly from their thoughtful and critical comments. They're also fun to grab drinks with.

I listened closely as my friends Saida, Jessica, and Heba described their expe-riences in writing, editing, and publishing their books. Hearing from y'all gave me confidence as I went through some of the same processes a year or so later. Thank you for being so open and honest about the joys and struggles of writ-ing, and for putting up with me texting you questions at random! Reuben Miller and Julian Go are two senior scholars who also shared their experience with writing books with me, and I'm grateful for those conversations and their wisdom and friendship. It's much easier to push forward when you have people who have done it the right way blazing the trail.

I owe special thanks to my colleague Jacob Groshek, whose influence, col-laboration, and friendship has shaped my trajectory. I was given access to training and tools for social media analytics through Jacob's lab and workshop. Most of our meetings took place on the basketball court, perhaps the epitome of work-life balance. I'm also thankful for my relationship with Natasha Polozenko, the brilliant designer, consultant, and illustrator who not only did the cover art and design, but also generously formatted all of the charts in this book. Natasha provides a constant model of diligence and hustle that I ascribe to match (though I know I never will). Thank you!

I finished the book during my first year as an associate professor at Columbia University, where I benefitted from departmental support and being around my new and wonderful colleagues. Heidi Allen and Melissa Begg supported my book workshop, which was instrumental in shaping the final manuscript.

On that note, I am grateful to the scholars who gave of their time and expertise to read this book at various stages, from Press reviewers and workshop participants to friends and colleagues willing to lend an eye. These include Julian Thompson, Robert Patterson, Marybeth Gasman, Eduardo Bonilla-Silva, Saida Grundy, and Jessie Daniels. No one has read more versions of chapters 3, 4, 5, and 7 than Julian. I cannot overstate the positive impact Julian has had on my work and life. Robert's wealth of knowledge and generous feedback is unmatched, and it seems I learn new words every time I read his work or have a conversation with him. Eduardo has a special place in the origins of this book and project, and I remain in awe that he would give of his time to help it develop. Similarly, Jessie was one of the first scholars on racism and technology that I found when I began reading on the subject in the early 2010s, and having Jessie's critical eye on an early version of this book was both an honor and opportunity to make this version much better. I am grateful to everyone who read some version of this – and I am in your debt.

I am thankful for my friend and colleague Desmond Patton; we began our first empirical research on social media together. Desmond took me out to lunch when I interviewed for the PhD program at the University of Chicago and has been giving me the inside scoop on academia ever since. I also have a hunch that I owe my book deal to Desmond—after he retweeted a paper I wrote, our shared editor Naomi Schneider sent me a note on Twitter asking if I had ever thought about writing a book on the subject. As fate would have it, I had just finished writing my book proposal, and I signed with California before I even sent the proposal out broadly. It felt like kismet.

I am grateful for my colleagues at Harvard University's Berkman Klein Center, who put up with me talking about this book project for a full year, provided feedback, and in general created an atmosphere where I was able to absorb their collective brilliance and knowledge of aspects of internet and technology studies that I had not been privy to. I regret that our Fellowship year was during the pandemic and that things weren't able to be in person, and still have hopes to meet up with more of y'all face to face.

I have been fortunate to have a number of talented and dedicated students working with me over the past five years. Luther Walls was the first, and is now an MD whom I can pester with informal medical questions. Sabriya Dillard, Pilar Haile-Damato, and Jamie Stepansky decided to be a part of the project as they were writing their Master's theses at Smith University, and without them I would not have been able to expand the project to the South, West, and Northeast. I am eternally grateful. Noor Toraif, my first PhD student, has been a dream mentee and coauthor, and breathed life into the Digital Race Lab. Many of my undergraduate students, including Rachel Chanderdatt, Khea Chang, and Maysa Whyte, have been an important part of the work coming out of the lab, much of which informed the way I have thought about this book. Portions of chapters 4 and 5 were published in *Social Problems* and *The Sociology of Race and Ethnicity* and are republished here with permission.

I am thankful for the brilliant Tamie Parker Song, my editor (and now friend) who spent several months reading and being in conversation with me about this book. Working with Tamie was an intense and surreal experience. I'm not sure what the book would have been without our time working together. Tamie consistently challenged me to think about things in ways I had not considered. The revisions for this book were closer to a rewrite, and my work with Tamie was an instrumental part of the process as I found my voice.

I am also grateful for my grad school mentors, without whom this project would never have gotten off the ground. Dr. William Sites, one of the most effective instructors I have ever encountered, and someone I hope to see on the basketball court before I retire from intramural sports. Dr. Gina Miranda Samuels, who gives an uncommon amount of time to students and student-led calls for change. I am part of the contingent of students who owes much to Gina. Dr. Cathy J. Cohen, who has shaped the way I think with her research, mentorship, and activism. Dr. Waldo E. Johnson Jr., whose friendship and mentorship extends beyond academia. I hope to model this style of relationship with students. And Dr. Charles Payne, who I followed from sociology to social work, and who has helped guide me through enemy territory with wisdom and critical instruction. I, like many other students, aspire to follow the Payne model of bridging the gap between the ivory tower and the communities we claim to serve.

When I wasn't writing at cafes in my home cities of Chicago, Boston, and New York, I wrote parts of this book in Carmel-by-the-Sea, Sedona, DC, Maine, and Lisbon. I also wrote a bit in Miami, and am greatly indebted to my dear friends David and Danielle who generously opened their home to me and allowed me to write on the porch, where the only distractions were the stray chicken that wandered into their yard by accident (and stayed for the blueberries), and the mango tree that kept calling to me, demanding I make sure its fruits did not go to waste.

This book starts with a story of me playing video games. These days, when I play video games, it is usually Fortnite with my kids (which I take just as seriously as I did Halo, back in the day, and am grateful that my kids give me this excuse to continue to be a gamer), or with The Black Justice League, a group of Black gamers (including my cousins CJ and David, of course, their little brother Joe, and my friend David, of Miami). I like to think we have institutionalized ways to avoid racism while gaming and have developed a community (or should I say, counterspace?) that extends beyond the internet.

And of course, I am grateful to the students in Los Angeles, Chicago, Atlanta, New York, and Boston who shared their time, experiences, and brilliance and without whom this book would have been neither possible nor interesting.

To the many students, faculty, friends, and family who have been my community—I thank you for investing in me. In the grand scheme of things, I'm new to academia, and this book is just the first act. But from where I'm sitting, it feels like my life's work. I've been thinking about and writing about these ideas for so long that it is a bit overwhelming to finally have them out in the world. Thank you for taking the time to read with me.

Appendix

All social media searches were conducted with the Brandwatch Consumer Research and social media analytics platform, which has its origins in academic research and provides full firehouse access to all undeleted tweets on Twitter, including past tweets.[1] Brandwatch allows for complex keyword searches and provides not only the numbers of tweets over time that match these searches, but can also give samples of tweets (either random, or based on some programmed characteristic) that can be used for more in-depth analysis. In both chapters 6 and 8, I used Brandwatch to gather the social media data. The syntax I used for each search is below.

I use an abductive grounded theoretical approach for the qualitative analyses in chapters 3, 4, 5, 6, and 7.[2] This modified form of grounded theory puts the same emphasis on memoing and the constant comparison between data and theory as does the original grounded theory, but does not seek to come to analysis free of any theoretical considerations.[3] Instead, this approach requires researchers to focus on findings that are surprising in light of existing theories, and use these surprising findings to develop new theory. This focus on surprising findings was central to this book project, as detailed in chapter 1, where I discuss the ways the project shifted its initial focus on online racism and its effects to its emphasis on innovative forms of technology-empowered resistance.

I conducted all interviews and focus groups in Chicago and Boston, but interviews and focus groups in Los Angeles were conducted by Pilar Haile-Damato, interviews in New York were conducted by Sabriya Dillard, and interviews in Atlanta were conducted by Jamie Stepansky and Sabriya Dillard. Pilar, Sabriya and Jamie were graduate students who worked on the study and used it for their MSW theses, and I am so grateful to have been able to work with them! There were a total of 86 students interviewed for this study, including 38 students in Chicago, 11 students in Los Angeles, 11 students in Atlanta, 10 students in New York, and 16 students in Boston. Interview quotes were minimally edited for readability, including removing excessive uses of the word "like" or "you know."

For all qualitative analysis, I went through multiple rounds of thematic coding, including open coding, focused coding, and theoretical coding. For chapters 3, 4, 5, and 7, I analyzed the data in the NVivo qualitative analysis program. For chapter 6, I analyzed the data in Microsoft Excel. Because large social media datasets from Brandwatch are downloaded in Excel format and include many data elements, including hyperlinks to each tweet, I found that analysis in Excel was more prudent than moving the data to NVivo. I created new columns in the data to represent the multiple rounds of coding, including columns for notes. Moreover, many tweets in chapter 6 required me to read *around the tweet,* taking context into account in order to fully understand its meaning. My analysis in chapter 6, therefore, includes the analysis of many tweets that are not in the dataset, but were adjacent to tweets in the dataset.

In keeping with Twitter API policy, I checked that none of the tweets I use in this dataset had been deleted as of the time of publication. As an additional measure of protection, and in keeping with my use of pseudonyms for interview participants, I removed names of the account holders behind the tweet examples shown in chapter 6. I also removed mentions of other Twitter users in the tweets for privacy.

In chapter 5 I referenced my "Online Racial Discussions Survey," which asks questions about internet and social media usage, engagement in racial discussions both in person and online, experiences with racism in person and online, political involvement, socio-demographic questions, and mental health outcomes. The survey was piloted with Boston University students and an online panel of college students from Qualtrics in 2019, and conducted through YouGov with a representative sample (using active sampling and statistical weighting) in 2021 and 2022. The sample includes 1,647 participants, 885 of whom were women, 742 men, 17 non-binary, and 3 another gender; 669 were White, 338 Black, 346 Hispanic, 232 Asian, 4 Native American, 34 multiracial, 23 other, and 1 Middle Eastern. Manuscripts from this research are being prepared for publication in peer-reviewed journals.

Statistical analyses were conducted in STATA and SAS. Tables 1 and 2 below show results of linear regression models that explore the effects of microaggressions on Kroenke, Spitzer, Williams, and Löwe's screening scale for depression and anxiety and Elo, Leppänen and Jahkola's single-item stress measure.[4] The types of microaggressions in Table 1, Table 3, and Table 4 include three subscales from a modified version of Keels, Durkee, and Hope's School-Based Racial and Ethnic Microaggressions Scale (subscales include inferiority, expectations of aggression, and stereotypical misrepresentations), and Table 1 also includes a single-item question that asks whether participants had experienced or witnessed microaggressions in the past year.[5] The types of microaggressions in Table 2 include three subscales from The Online Microaggressions Scale, an original scale modified from Keels, Durkee, and Hope (subscales include inferiority, expectations of aggression, and stereotypical misrepresentations), and a single-item question that asks whether participants had experienced or witnessed online microaggressions in the past year.

Other regressions were conducted to determine whether resisting racism moderated the effects of racism on mental health outcomes. Tables 3, 4, and 5 show the moderation effect of three types of resistance, including the frequency of responding to racism (Table 3), the frequency of witnessing a response to racism (Table 4), and whether a participant has posted online about an in-person incident of racism in the year preceding their participation in the survey (Table 5). These tables report the moderation effect (interaction term) for eighteen statistical models, each of which also included the same covariates as the models in tables 1 and 2 (ie age, gender, income, time on social media, and education). These results suggest that increased resistance (as formulated above) reduces the harmful relationship between experiencing racism or microaggressions on mental health.

TABLE I

Linear regression models of in-person microaggressions (IV) and health outcomes (DV)

Variables		Depression and anxiety				Stress			
		Inferiority	Aggression	Stereotypes	Single-item	Inferiority	Aggression	Stereotypes	Single-item
Microaggression		−0.4838***	−0.3982***	−0.4515***	0.2311***	−0.6838***	−0.5324***	−0.6637***	0.3384***
Time on social media		−0.0671***	−0.0721***	−0.0715***	−0.0724***	−0.0742***	−0.0824***	−0.0796***	−0.081***
Race	Asian	0.3042***	0.1973***	0.309***	0.2079***	0.2928***	0.1411	0.3063***	0.1576
	Black	0.3783***	0.3963***	0.394***	0.2682***	0.523***	0.5334***	0.5562***	0.3709***
	Hispanic	0.2276***	0.2134***	0.2361***	0.1378**	0.2837***	0.2561***	0.3024***	0.1579*
	White	0	0	0	0	0	0	0	0
Gender	Woman	−0.0847*	−0.1102**	−0.0677	−0.1082**	−0.2428***	−0.276***	−0.2185***	−0.2778***
	Man	0	0	0	0	0	0	0	0
Family income		0.0225**	0.024***	0.0249***	0.0243***	0.0246**	0.0267**	0.0281**	0.0273**
Education	2-year	0.0045	−0.0135	−0.05	0.0109	−0.0972	−0.1222	−0.1766	−0.0871
	4-year	0.0696	0.0555	0.0452	0.058	−0.0217	−0.0419	−0.0568	−0.0379
	HS graduate	0.0301	0.0356	−0.0089	0.0309	0.0848	0.0933	0.0267	0.0852
	No HS	−0.0981	−0.0739	−0.1355	−0.084	0.0722	0.1074	0.0151	0.091
	Post-grad	−0.1095	−0.1313	−0.1349	−0.0892	−0.1925	−0.2257*	−0.2269*	−0.1601
	Some college	0	0	0	0	0	0	0	0
Age		0.0086***	0.0092***	0.009***	0.0076***	0.0139***	0.0149***	0.0144***	0.0124***

* Indicates significance at the .1 level.

** Indicates significance at the .05 level.

*** Indicates significance at the .01 level.

TABLE 2

Linear regression models of online microaggressions (IV) and health outcomes (DV)

Variables		Depression and anxiety				Stress			
		Inferiority	Aggression	Stereotypes	Single-item	Inferiority	Aggression	Stereotypes	Single-item
Microaggression		−0.3871***	−0.3533***	−0.3862***	0.1919***	−0.4988***	−0.4389***	−0.593***	0.2949***
Time on social media		−0.0672***	−0.0701***	−0.0718***	−0.0761***	−0.0767***	−0.0812***	−0.0792***	−0.0856***
Race	Asian	0.2322***	0.1903***	0.2726***	0.1562**	0.1859*	0.1319	0.2581**	0.0794
	Black	0.2777***	0.2771***	0.3102***	0.1996***	0.3709***	0.3665***	0.4401***	0.2705***
	Hispanic	0.172***	0.1667***	0.1927***	0.0937	0.1979**	0.1888**	0.2437***	0.0916
	White	0	0	0	0	0	0	0	0
Gender	Woman	−0.1046**	−0.0975**	−0.1033**	−0.1225***	−0.267***	−0.2568***	−0.2728***	−0.3023***
	Man	0	0	0	0	0	0	0	0
Family income		0.0228***	0.0235***	0.0269***	0.0262***	0.0253**	0.0261**	0.0312***	0.0301***
Education	2-year	−0.0064	0.0029	−0.0019	0.0114	−0.1128	−0.1013	−0.1055	−0.0851
	4-year	0.0647	0.0754	0.0738	0.0499	−0.0302	−0.0174	−0.0134	−0.05
	HS graduate	0.037	0.0449	0.0347	0.0177	0.0956	0.1058	0.0901	0.0639
	No HS	−0.1007	−0.0781	−0.1001	−0.1009	0.0735	0.1034	0.0647	0.0634
	Post-grad	−0.1256	−0.1301	−0.1208	−0.1351	−0.22	−0.2274*	−0.2033	−0.2252*
	Some college	0	0	0	0	0	0	0	0
Age		0.0087***	0.0094***	0.0086***	0.0082***	0.0143***	0.0153***	0.0137***	0.0131***

* Indicates significance at the .1 level.

** Indicates significance at the .05 level.

*** Indicates significance at the .01 level.

TABLE 3

Moderation effect: Frequency of responding to racism

Outcome	Microaggression subscales	White		Non-white	
		Estimate	p-value	Estimate	p-value
Depression/ anxiety	Inferiority	0.111	0.436	−0.156	0.021*
	Aggression	0.039	0.789	−0.066	0.284
	Stereotypical misrepresentations	0.046	0.722	−0.079	0.289
Stress	Inferiority	0.303	0.126	−0.357	0.000***
	Aggression	0.121	0.550	−0.192	0.033*
	Stereotypical misrepresentations	0.100	0.578	−0.333	0.002**

TABLE 4

Moderation effect: Frequency of witnessing responses to racism

Outcome	Microaggression subscales	White		Non-white	
		Estimate	p-value	Estimate	p-value
Depression/ anxiety	Inferiority	−0.078	0.610	−0.188	0.012*
	Aggression	−0.126	0.466	−0.144	0.037*
	Stereotypical misrepresentations	−0.095	0.506	−0.161	0.069
Stress	Inferiority	0.015	0.945	−0.415	0.000***
	Aggression	0.093	0.702	−0.194	0.055
	Stereotypical misrepresentations	−0.146	0.466	−0.353	0.005**

TABLE 5
Moderation effect: Posts online about the experience

Outcome	Microaggression subscales	White		Non-white	
		Estimate	p-value	Estimate	p-value
Depression/ anxiety	Inferiority	−0.091	0.859	−0.695	0.006**
	Aggression	−0.279	0.694	−0.633	0.004**
	Stereotypical misrepresentations	0.038	0.934	−0.976	0.000***
Stress	Inferiority	−0.333	0.650	−0.458	0.237
	Aggression	−0.852	0.404	−0.670	0.052
	Stereotypical misrepresentations	−0.461	0.485	−0.827	0.034*

BRANDWATCH SYNTAX

Chapter 6: Double-sided Consciousness
- Figure 1
  - Syntax: Wypipo

Chapter 8: Racism is Trending
- Figure 2. Total Tweets on Twitter by Month, 2011–2021
  - Syntax: site: Twitter.com
- Figure 3. Monthly Mentions of Racism or Racist on Twitter, 2011–2021
  - Syntax: racism OR racist
- Figure 4. Monthly Mentions of Structural, Systemic, or Institutional Racism on Twitter, 2011–2021
  - Syntax: (racism AND (structural OR institutional OR systemic)) OR (racist AND (structures OR institutions OR systems))
- Figure E. Monthly Mentions of Microaggressions on Twitter, 2011–2021
  - Syntax: (microaggression OR microaggressions OR microaggressive OR micro-aggression OR micro-aggressions OR micro-aggressive OR "micro aggression" OR "micro aggressions" OR "micro aggressive")
- Figure F. Monthly Mentions of White Privilege on Twitter, 2011–2021
  - Syntax: "white privilege" OR whiteprivilege
- Figure G. Monthly Mentions of Cultural Appropriation on Twitter, 2011–2021
  - Syntax: ((culture OR cultural) AND (appropriate OR appropriating OR appropriation OR appropriated)) OR culturalappropriation
- Figure H. Monthly Mentions of Intersectionality on Twitter, 2011–2021
  - Syntax: intersectionality OR intersectional
- Figure I. Monthly Mentions of Patriarchy on Twitter, 2011–2021
  - Syntax: patriarchal OR patriarchy OR patriarchies
- Figure J. Monthly Mentions of Heteropatriarchy on Twitter, 2011–2021
  - Syntax: heteropatriarchal OR heteropatriarchy OR heteropatriarchies OR "hetero patriarchal" OR "hetero patriarchy" OR "hetero patriarchies"
- Figure K. Monthly Mentions of Dog Whistle on Twitter, 2011–2021
  - Syntax: dogwhistle OR "dog whistle"

- Figure L. Monthly Mentions of Vote Suppression on Twitter, 2011–2021
  - Syntax: (vote OR voter OR voting OR voters) AND (repression OR repress OR repressed OR suppression OR suppress OR suppressed)
- Figure M. Monthly Mentions of Abolition on Twitter, 2011–2021
  - Syntax: abolition OR abolitionist
- Figure N. Monthly Mentions of Defund/Abolish Police on Twitter, 2011–2021
  - Syntax: Police AND (defund OR abolish)
- Figure O. Monthly Mentions of Abolish ICE on Twitter, 2011–2021
  - Syntax: ICE AND (defund OR abolish OR shutdown)
- Figure P. Monthly Mentions of Reparations on Twitter, 2011–2021
  - Syntax: reparation OR reparations
- Figure Q. Monthly Mentions of Juneteenth on Twitter, 2011–2021
  - Syntax: Juneteenth
- Figure R. Monthly Mentions of Black Lives Matter on Twitter, 2011–2021
  - Syntax: Blacklivesmatter OR "Black lives matter"
- Figure S. Monthly Mentions of All Lives Matter on Twitter, 2011–2021
  - Syntax: alllivesmatter OR "all lives matter"
- Figure T. Monthly Mentions of Model Minority on Twitter, 2011–2021
  - Syntax: modelminoritymyth OR "model minority" OR modelminority OR "model minority myth" OR "model minorities" OR modelminorities
- Figure U. Monthly Mentions of Palestine on Twitter, 2011–2021
  - Syntax: Palestine

Notes

1. AN INTELLECTUAL PUZZLE

1. Sue, *Microaggressions in Everyday Life*.

2. Bonilla-Silva, *Racism without Racists*.

3. After the advances Black folks made post-emancipation (the Reconstruction period), Jim Crow laws enforced legal segregation and unequal treatment of Black people beginning in the late nineteenth century and lasting until the landmark legal changes brought on by the victories of the civil rights movement, including the Civil Rights Act of 1964, the Voting Rights Act of 1965, and the Fair Housing Act of 1968.

4. Christopherson, "Positive and Negative Implications of Anonymity"; Lapidot-Lefler and Barak, "Effects of Anonymity."

5. Gray, "Intersecting Oppressions and Online Communities"; Gray, *Intersectional Tech*.

6. See, e.g., Daniels, *Cyber Racism;* Hughey and Daniels, "Racist Comments at Online News Sites."

7. Hughes et al., "Parents' Ethnic-Racial Socialization Practices."

8. Jurgensen, "Digital Dualism versus Augmented Reality."

9. Delgado and Stefancic, *Critical Race Theory: The Cutting Edge*; Delgado and Stefancic, *Critical Race Theory: An Introduction*.

10. Small, "'How Many Cases Do I Need?'"

11. Surprising findings are central to the abductive grounded theoretical approach to qualitative data analysis, something I discuss more in the appendix.

12. Glaser and Strauss, *Discovery of Grounded Theory*; Timmermans and Tavory, "Theory Construction in Qualitative Research."

13. Eschmann, "How the Internet Shapes Racial Discourse."; Eschmann, "Unmasking Racism"; Eschmann, "Digital Resistance."

14. Auxier and Anderson, "Social Media Use in 2021."

15. Brock, *Distributed Blackness*, 81.

16. Cottom, "Black Twitter Is Not a Place."

17. Hill, "'Thank You, Black Twitter,'" 287.

18. Jackson et al., *#HashtagActivism*; Brown et al., "#SayHerName"; Ray et al., "Ferguson and the Death of Michael Brown."

19. This analogy, moving from a pond to an ocean, comes from my friend and colleague Saida Grundy.

20. "Ku Klux Klan," Southern Poverty Law Center.

21. Of course, openly White supremacist groups never fully went away (see Daniels 2009), but they are largely seen as extremist organizations and represent outliers among racists, not the "normative" racism that is most commonly experienced by people of Color (see Bonilla-Silva 2017), which I'll discuss more in the next chapter. These groups have worked hard to enter the mainstream (see Daniels 2018) and attempted to separate themselves from their violent, terrorist roots, as their existence today is dependent on constitutional free speech, not community fear that they might engage in acts of violence with legal impunity.

22. Williams, "How Racism Makes Us Sick."

23. "You will not replace us" was a rallying cry at the Unite the Right White Supremacist rally in Charlottesville, Virginia, 2017. See Heim, "Recounting a Day of Rage, Hate."

24. Haltiwanger, "Trump's History of Support."

25. Yam, "Anti-Asian Hate Crimes."

26. Wu, "Cory Booker Denounces White Supremacy"; Levenson et al., "Mass Shooting at Buffalo Supermarket."

27. Yousef, "31 Members of a White Nationalist Group."
28. Daniels, "Algorithmic Rise of the 'Alt-Right.'"
29. Rascoe, "Why Are White Nationalist Groups."

2. ONCE WE WERE COLORBLIND

1. LaPiere, "'Attitudes vs. Actions.'"
2. May, "Velvet Rope Racism."
3. Dwyer, "Donald Trump: 'I Could . . .'"
4. "Cuomo: What Would You Do If Trump Said."
5. "Cuomo to Trump Supporter."
6. Domonoske, "Trump Fails to Condemn KKK."
7. Reilly, "Donald Trump Blames 'Both Sides.'"
8. Timberg and Dwoskin, "Proud Boys, Right-Wing Extremists."
9. Amatulli, "Trump Says He's 'The Least Racist Person.'"
10. Krysan and Moberg, "Trends in Racial Attitudes."
11. Abel, *Signs of the Times,* 170–71.
12. Jeffrey and Bhattacharjee, "Photos Show Violent Clashes"; Jacobo, "Visual Timeline on How the Attack."
13. Broadwater and Schmidt, "Trump Urged Armed Supporters to Capitol."
14. Bobo, "Racism in Trump's America," S88. Wilson, *Power, Racism and Privilege.*
15. Bonilla-Silva, *Racism without Racists.*
16. Corasaniti, "Georgia G.O.P. Passes Major Law"; Jones, "Texas Senate Joins GOP Voter Suppression Push."
17. Herron, "Bill That Made Tempers Rise"; Meckler and Rabinowitz, "Lines That Divide."
18. Apple, "Use Face ID While Wearing a Mask."
19. Grimm, "Social Desirability Bias."
20. McConahay, "Modern Racism, Ambivalence."
21. Kurzban et al., "Can Race Be Erased?"
22. Eagly and Diekman, "What Is the Problem?," 20.
23. Correll et al., "Across the Thin Blue Line," 1006.
24. Correll et al., "Police Officer's Dilemma."

25. FBI Counterterrorism Division, "(U) White Supremacist Infiltration"; Downs, "FBI Warned of White Supremacists"; Tobar, "Deputies in 'Neo-Nazi' Gang."

26. Singal, "Psychology's Favorite Tool for Measuring Racism."

27. Picca and Feagin, *Two-Faced Racism.*

28. Bonilla-Silva, "Structure of Racism in Color-Blind," 1359.

29. Rothstein, *Color of Law;* DeWitt, "Decision to Exclude Agricultural and Domestic Workers."

30. Kil, "Fearing Yellow, Imagining White."

31. Frankenberg et al., "Harming Our Common Future."

32. Alexander, *New Jim Crow.*

33. Crimmins and Saito, "Trends in Healthy Life Expectancy"; Richardson and Norris, "Access to Health and Health Care"; Alegria et al., "Disparities in Treatment for Substance Use Disorders"; Prather et al., "Impact of Racism on the Sexual and Reproductive Health"; Trent et al., "Impact of Racism on Child and Adolescent Health"; Black et al., " Relationship between Perceived Racism/ Discrimination."; Brondolo et al., "Racism and Hypertension"; Fernando, "Racism as a Cause of Depression"; Pieterse et al., "Perceived Racism and Mental Health," 1.

34. Bobo et al., "Laissez-Faire Racism."

35. Chavez, *Latino Threat.*

36. Massey and Pren, "Origins of the New Latino Underclass."

37. Massey and Pren.

38. Blake, "No, Biden's New Border Move"; Rose, "Inquiry into Border Agents on Horseback Continues."

39. Bonilla-Silva, *Racism without Racists.*

40. Bonilla-Silva, *Racism without Racists* (6th 2021 edition), 2.

41. Nellis, "Color of Justice."

42. Steen et al., "Images of Danger and Culpability"; Petersilia, "Racial Disparities in the Criminal Justice System"; Everett and Wojtkiewicz, "Difference, Disparity, and Race/Ethnic Bias."

43. Schulman et al., "Effect of Race and Sex on Physicians' Recommendations."

44. Gregory et al., "Achievement Gap and the Discipline Gap."

45. Kang, "Cyber-Race."

46. West and Thakore, "Racial Exclusion in the Online World."

47. Kafai et al., "'Blacks Deserve Bodies Too!'"

48. Nakamura, *Digitizing Race.*

49. Bots are robot accounts, or those programmed to spam users with pre-written responses who tweet about certain topics.

50. Rashid, "Emergence of the White Troll."

51. Andrew, "Twitter Suspends Fake Accounts."

52. Eustachewich, "White Male New Hampshire Professor."

53. Christopherson, "Positive and Negative Implications of Anonymity."

54. Groshek and Cutino, "Meaner on Mobile," 5.

55. Lapidot-Lefler and Barak, "Effects of Anonymity."

56. Bartlett et al., *Anti-social Media.*

57. I think some graffiti can make trains or brick walls look better, and represents one of the key elements of hip hop culture. For the purpose of analogy here, I'm thinking less about graffiti as art, and more the vandalism variety.

58. Stephens-Davidowitz, "Cost of Racial Animus."

59. Daniels, "Cloaked Websites."

60. I don't recommend navigating to this site, because when I last tried it in 2021 I was prompted to install something on my computer (possibly some type of malware). For more details about what the site used to be, see the above article by Daniels.

61. Noble, *Algorithms of Oppression;* Benjamin, *Race after Technology.*

62. Benjamin.

63. Boyd, *It's Complicated.*

64. Meraz, "Is There an Elite Hold?"

65. Brock, "Race Matters."

66. Richardson, "Bearing Witness While Black"; Richardson, *Bearing Witness While Black.*

67. Folayan and Davis, *Whose Streets?*

68. Payne, *I've Got the Light of Freedom.*

69. Payne, 52, 99.

70. Hall, "Long Civil Rights Movement."

71. Hall.

72. Groshek and Tandoc, "Affordance Effect."

73. Anderson, "History of the Hashtag #BlackLivesMatter."

74. Ray et al., "Ferguson and the Death of Michael Brown"; Hill, "'Thank You, Black Twitter.'"

75. Jackson et al., *#HashtagActivism;* Noble and Brendesha Tynes, *Intersectional Internet;* Brown et al., "#SayHerName."

76. Williams, "Black Memes Matter."

77. Cottom, "Where Platform Capitalism and Racial Capitalism Meet."

78. Brock, *Distributed Blackness,* 6.

79. An abbreviation used to refer to Black, Indigenous, and people of Color.

3. MASK ON: RULES OF RACIAL ENGAGEMENT

1. Delgado and Stefancic, *Critical Race Theory: An Introduction.*

2. Kelly, "Trump Bars 'Propaganda' Training Sessions."

3. Schwartz, "Map: Where Critical Race Theory."

4. "Unsolved Mysteries," The Daily Show with Trevor Noah.

5. Depenbrock, "Federal Judge Finds Racism."

6. Palos, *Precious Knowledge.*

7. Kendi, *Stamped from the Beginning.*

8. Morgan, "George Washington and the Problem of Slavery."

9. Smith, *How the Word Is Passed.*

10. Timm, "Trump Says Obama Didn't Reform Policing."

11. Baker, "'Murder Hornets' in the U.S.'"

12. *The Good Wife,* season 1, episode 10, "Lifeguard," directed by Paris Barclay, written by Tom Smutts, starring Julianna Margulies, aired December 15, 2009, on CBS.

13. Simes, *Punishing Places.*

14. Alexander, *New Jim Crow.*

15. Yosso et al., "Critical Race Theory, Racial Microaggressions."

16. Parent, "Liberal Legacy," 8.

17. Nunnally and Carter, "Moving from Victims to Victors."

18. Binkley, "Harvard Pledges $100 Million."

19. Solorzano et al., "Critical Race Theory, Racial Microaggressions"; Yosso et al., "Critical Race Theory, Racial Microaggressions."

20. Hermann et al., "Pro-Trump Rally Descends into Chaos."

21. Jacobo, "Visual Timeline on How the Attack."

22. Hall et al., "At Least 865 People Have Been Charged."

23. Tolan, "DC Police Made Far More Arrests."

24. "What Does Free Speech Mean?," United States Courts.

25. Loury, "Self-Censorship in Public Discourse."

26. Bush, "Brief History of PC."

27. Banning, "Limits of PC Discourse."

28. Bobo et al., "Laissez-Faire Racism."

29. Carlisle, "Can We Talk?"

30. Delgado, "Legal Realism," 275.

31. Leonardo and Zembylas, "Whiteness as Technology of Affect."

32. Kendi, *Stamped from the Beginning*; Kendi, *How to Be an Antiracist*.

33. I teach in a school of Social Work, where justice and care for the marginalized is central to the discipline, so my students may or may not be typical when compared with students in some other disciplines or departments.

34. A part-time grad-student university official that supervises RAs and is responsible for putting on events and responding to emergencies and incidents of sexual assault or bias.

35. Lopez, "Ta-Nehisi Coates Has an Incredibly Clear Explanation."

36. Lipsky, *Street-Level Bureaucracy*.

37. Kennedy, "Who Can Say 'Nigger'?"

38. A rap artist from Chicago.

39. Twain, *The Adventures of Tom Sawyer*.

40. Lorde, "Uses of Anger."

41. Jaschick, "The Chicago Letter and Its Aftermath."

42. Stone, "Aims of Education Address 2016."

43. Faust, "Harvard University 2017 Commencement Speech"; Goldberg, "Speech, Protest and the Rules"; Eisgruber, "Op-ed by Princeton President Eisgruber."

44. Jaschick, "Incidents Roil Campuses"; Yosso et al., "Critical Race Theory, Racial Microaggressions."

45. DiAngelo, "White Fragility."

46. Lorde, "Uses of Anger," 283.

4. MASK OFF: REVELATIONS AND NEW REALITIES

1. Pollock, *Colormute.*
2. Carter, *From George Wallace to Newt Gingrich.*
3. Hawkins et al., "Hidden Tribes."
4. Tynes et al., "Online Racial Discrimination."
5. Bartlett et al., *Anti-social Media.*
6. Miller, "Sorority Sister Speaks Out"; Cook and Bensen, "Racist Photo Sparks Outrage"; Ellis, "Former Kentucky Police Official."
7. Lee and Park, "In 15 High-Profile Cases."
8. Brock, "Critical Technocultural Discourse Analysis."
9. Bonilla-Silva, *Racism without Racists.*
10. Picca and Feagin, *Two-Faced Racism.*
11. Both Daniels 2013 and Boyd 2014 have written about this unique conflation of public and private space.
12. Tripodi, "Yakking about College Life."
13. This quote refers to the tragic death of two Chinese students on a plane crash from South Korea to San Francisco, an event that was in the news shortly before the comment was published.
14. Cabrera, "'But I'm Oppressed Too.'"
15. Daniels, "Race and Racism in Internet Studies."
16. Márquez, "Black Mohicans."
17. Duster, "Diversity Project."
18. "Personal Privilege Profile," Interaction Institute for Social Change.
19. Joravsky, "Did the NBA Blacklist Former Chicago Bulls Player"; Washington, "We Finally Have Answers about Michael Jordan."
20. As a Chicago kid I can't help but be a Jordan fan, and I still feel like I have superpowers when I put on a pair of Jordan brand shoes.
21. Jordan, "Michael Jordan."
22. It wasn't until June of 2020, at the height of the Black Lives Matter protests and uprisings, that the Jordan brand put out a statement that said unequivocally that Black lives matter, and announced plans for $100 million in donations to racial equity programs over 10 years. This statement was not from Jordan himself, and pointed out that the brand was bigger than one person (Jordan).

23. Kamenetz et al., "Harvard Rescinds Admission Of 10 Students."

24. Bonilla-Silva, "Linguistics of Color Blind Racism," 43.

25. Jaschick, "U of Chicago Letter to New Students"; Kamenetz et al., "Harvard Rescinds Admission Of 10 Students."

26. Rowan-Kenyon et al., "Social Media in Higher Education."

27. Herron, "I Used to be a 911 Dispatcher."

28. Siegel, "Two Black Men Arrested at Starbucks Settle"; Herreria, "Woman Calls Police"; Perez, "White Man Calls Police."

29. Williams, "Black Memes Matter."

5. DIGITAL RESISTANCE

1. The original says "man," but I edited the quote for gender-neutral language.

2. Located in the heart of Chicago's South Side, security at UChicago is something that is taken very seriously—and, unfortunately, has historically been racialized.

3. This is commonly known as "the Black handshake." For a reference, see how President Obama shakes basketball player Kevin Durant's hand here (https://www.youtube.com/watch?v=LmnqihRlcaI&ab_channel=MyBestVines), or the Key and Peele comedy sketch referencing Obama's selective use of the handshake here (https://www.youtube.com/watch?v=nopWOC4SRm4&ab_channel=ComedyCentralComedyCentralVerified).

4. Bonilla-Silva writes about how the New Racism, marked by subtle discrimination and little open racial discourse, is actually harder to combat. This is one reason why *unmasking racism* is so important!

5. Pierce, "Offensive Mechanisms."

6. Eschmann et al., "Making a Microaggression."

7. Sue, *Microaggressions in Everyday Life.*

8. Pyke, "What Is Internalized Racial Oppression."

9. Sue, *Microaggressions in Everyday Life.*

10. As I mention in chapter 1, one of the tenets of critical race theory is to focus on the stories of marginalized peoples, not only because they may be more sensitive to the mechanisms of oppression, but also because their experi-

ences are typically ignored by folks in power and not acknowledged by official gatekeepers of knowledge.

11. Solorzano et al., "Critical Race Theory, Racial Microaggressions"; Yosso et al., "Critical Race Theory, Racial Microaggressions."

12. Eschmann, "How the Internet Shapes Racial Discourse"; Eschmann, "Digital Resistance," 267.

13. Yosso et al., "Critical Race Theory, Racial Microaggressions"; Solorzano et al., "Critical Race Theory, Racial Microaggressions"; Case and Hunter, "Counterspaces"; Jackson et al., *#HashtagActivism;* Hill, "'Thank You, Black Twitter.'"

14. Moffatt, *Coming of Age in New Jersey.*

15. *Uncle Tom* is a term used to describe Black folks who are seen as being submissive to Whites.

16. See the appendix for more details around these findings, the survey sample, the measures used, and regression tables.

17. Williams et al., "Racial Differences in Physical and Mental Health"; Pieterse et al., "Perceived Racism and Mental Health"; Williams and Collins, "Racial Residential Segregation."

18. Brondolo et al., "Racism and Hypertension"; Fernando, "Racism as a Cause of Depression"; Pieterse and Carter, "Role of Racial Identity in Perceived Racism"; Graham et al., "Mediating Role of Internalized Racism," 369; Blume et al., "Relationship of Microaggressions with Alcohol Use"; Keels et al., "Psychological and Academic Costs"; Nadal et al., "Adverse Impact of Racial Microaggressions"; Eschmann et al., "Context Matters."

19. hooks, *Yearning.*

20. Steele and Aronson, "Stereotype Threat and the Intellectual Test Performance."

21. Spencer et al., "Stereotype Threat and Women's Math Performance."

22. Pager et al., "Discrimination in a Low-Wage Labor Market."

23. Schulman et al., "Effect of Race and Sex on Physicians' Recommendations."

24. Fryer, "Empirical Analysis of Racial Differences"; Ross et al. "Racial Disparities in Police Use"; Kramer and Remster, "Stop, Frisk, and Assault?"; Darrell Steffensmeier and Demuth, "Ethnicity and Sentencing Outcomes"; Petersilia, "Racial Disparities in the Criminal Justice System"; Steen et al.,

"Images of Danger and Culpability"; Lundman and Kaufman, "Driving While Black."

25. Ture and Hamilton, *Black Power.*

26. Eschmann, "Digital Resistance," 267.

27. Kawakami et al., "Mispredicting Affective and Behavioral Responses."

28. DiAngelo, "White Fragility."

29. Evans and Moore, "Impossible Burdens:"; Solorzano et al., "Critical Race Theory, Racial Microaggressions"; Pérez Huber and Solorzano, "Racial Microaggressions as a Tool."

30. See the appendix for more details around these findings, the survey sample, the measures used, and regression tables.

31. While Hochman includes racial ideologies as a part of a list of conservative concerns (i.e., CRT bans as victories), discussions of race and racism are not the only online conversations that might "radicalize" the Right.

32. Eschmann et al., "Bigger than Sports."

33. Eschmann et al., 5.

6. DOUBLE-SIDED CONSCIOUSNESS

1. When writing this, I searched YouTube looking for the video and found numerous examples of similar videos with crowds chanting the n-word with Kendrick.

2. Stolworthy, "Kendrick Lamar Explains Why He Criticised."

3. Du Bois, *Souls of Black Folk,* 16.

4. Thomas, "Du Bois, Double Consciousness"; Martinez, "Double-Consciousness of Du Bois"; Falcón, "Mestiza Double Consciousness:"; Wang, "'Double Consciousness,' Sociological Imagination"; Islam, "Muslim American Double Consciousness."

5. Marable and Agard-Jones 2008 suggest that this line is between North America, Europe, and Japan on the side of the colonizers, indicating that it is certainly possible for non-White countries to engage in economic oppression and colonization. Still, the vast majority of colonization took place at the hands of European (and later North American) White-dominated global powers.

6. Marable and Agard-Jones, *Transnational Blackness.*

7. Steele and Aronson, "Stereotype Threat and the Intellectual Test Performance."

8. Sellers and Shelton, "Role of Racial Identity in Perceived Racial Discrimination."

9. Sue, *Microaggressions in Everyday Life*.

10. Klein, "Trump Said 'Blame on Both Sides'"; Nagel, "Biden Cites Charlottesville"; Vagianos, "Jemele Hill on Calling Trump"

11. Cottom, "Black Twitter Is Not a Place."

12. Brock, *Distributed Blackness*, 32.

13. Cohen, "Deviance as Resistance."

14. Cottom, "Black Twitter Is Not a Place."

15. I speak about this more in chapter 3.

16. Yosso et al., "Critical Race Theory, Racial Microaggressions."

17. In tweets that mentioned another user (which is done by using the @ symbol followed by a user or profile name, i.e., @user) I have removed the name being mentioned in order to protect the privacy of the users mentioned by these tweets.

18. Black, "Fanon and DuBoisian Double Consciousness," 14.

19. Gates and West, *Future of the Race*.

20. Levine, *Black Culture and Black Consciousness*.

21. This refers to a music award for best hip hop album being given to a White artist (Macklemore) over the incomparable Kendrick Lamar. Even Macklemore admitted Kendrick should have won the award in a private text message he sent Kendrick, which Macklemore posted on social media here: https://www.instagram.com/p/jqXYYAwK_y/.

22. "Do You Think Bill Clinton."

23. Binoya, "Racism in Porn Industry"; Snow, "Black Porn Stars Come Forward"; Snow, "Rise of Racist Porn."

24. Sellers et al., "Multidimensional Model of Racial Identity"

25. Hoggard et al., "Racial Cues and Racial Identity Implications"; Sellers et al., "Racial Identity Matters."

26. Hughes et al., "Parents' Ethnic-Racial Socialization Practices."

27. Dazey, "Rethinking Respectability Politics."

28. Cohen, "Deviance as Resistance," 41.

29. Smith, *Racial Battle Fatigue in Higher Education*.

30. Jones, "#BlackLivesMatter," 105.

31. Cottom, "Black Twitter Is Not a Place."

32. Cohen, "Deviance as Resistance," 38.

33. Cohen, 38.

7. PROTEST, POSTERS, AND QR CODES

1. Davis, *Are Prisons Obsolete?*

2. Eschmann et al., "Tweeting toward Transformation."

3. Moorwood, "Twitter Reacts to Stacey Dash's Apology."

4. We talk about this incident in more detail in chapter 3.

5. It seems CRT has become the catchphrase conservatives use to refer to any antiracist or diversity elements of a school curriculum.

6. Cogburn et al., *1000 Cut Journey.*

7. The square, dot-filled image we scan with our phone's camera to open up a digital menu at restaurants that put the paper menus away during and after the Covid pandemic (QR codes are not restricted to restaurant settings, however).

8. Jackson et al., *#HashtagActivism.*

9. Ransby, *Ella Baker and the Black Freedom Movement;* Payne, *I've Got the Light of Freedom.*

10. Payne, *I've Got the Light of Freedom,* 93.

11. Payne.

12. Hall, "Long Civil Rights Movement."

13. Smock, *Democracy in Action.*

14. Organizers did use social media to plan their meetings and strategies, but the actual resistance took place in-person with traditional methods such as drafting letters, meeting with the administration, and holding a forum.

15. Leonardo and Porter, "Pedagogy of Fear"

16. Viala-Gaudefroy and Lindaman, "Donald Trump's 'Chinese Virus.'"

17. Jackson et al., *#HashtagActivism.*

18. Ray et al., "Ferguson and the Death of Michael Brown."

19. Cohen and Kahne, "Participatory Politics," 8.

20. Rundle et al., "Doing Civics in the Digital Age."

21. Cohen, "Reimagining Political Participation."

22. Bond et al., "A 61-Million-Person Experiment."

23. Gladwell, "Small Change"; Kristofferson et al., "Nature of Slacktivism."

24. Cohen, "Reimagining Political Participation"; Boulianne, "Social Media Use and Participation"; Marcia Mundt et al., "Scaling Social Movements"; Lane et al., "From Online Disagreement to Offline Action"; Conroy et al., "Facebook and Political Engagement."

25. Goudie et al., "Chicago Has Authorized Nearly $67M."

26. "Settling For Misconduct."

27. Groshek and Tandoc, "Affordance Effect."

28. Barberá et al., "Tweeting from Left to Right"; Colleoni et al. "Echo Chamber or Public Sphere?"

29. As indicated by the brackets, I changed the name of the group for privacy purposes.

30. Peebles, "7 Common Sense Reasons."

31. Smock, *Democracy in Action.*

8. RACISM IS TRENDING

1. Anderson, *White Rage.*

2. Schaffner et al., "Explaining White Polarization in the 2016 Vote"; Rhodes, "Voting Rights Politics in the Age of Obama."

3. Sela et al., "Changes in the Discourse of Online Hate Blogs"; Potok, "'Patriot' Movement Explodes."

4. Ryan, "'This Was a Whitelash.'"

5. Jones and Kludt, "Barack Obama to Black Voters."

6. Berenson, "David Duke Says He and Donald Trump."

7. "Presidential Approval Ratings—Donald Trump," Gallup.

8. Montanaro, "President-Elect Joe Biden."

9. Brandom, "Why Platforms Had to Cut off Trump."; Alba et al., "What Happened When Trump."

10. Nobles, "Marjorie Taylor Greene."

11. Bond, "Elon Musk Says."

12. Uebele, "All Rev'd Up."

13. This pattern holds for mentions of racist, racism, institutional racism, and systemic racism.

14. Eschmann et al., "Making a Microaggression."

15. A great example of teaching folks how to think about the ways they experience privilege: McIntosh, *White Privilege.*

16. Rogers, "From Cultural Exchange to Transculturation."

17. Jones, *Reclaiming Our Space;* Steele, *Digital Black Feminism.*

18. Bailey, "On Misogynoir"; Bailey, *Misogynoir Transformed.*

19. Crenshaw, "Mapping the Margins."

20. Combahee River Collective, "Combahee River Collective Statement."

21. Payne, *I've Got the Light of Freedom;* Levine, *Bayard Rustin and the Civil Rights Movement.*

22. "Black Lives Matter: About," Black Lives Matter.

23. Albertson, "Dog-Whistle Politics."

24. Corasaniti, "Georgia G.O.P. Passes Major Law"; Jones, "Texas Senate Joins GOP Voter Suppression Push."

25. Davis, *Are Prisons Obsolete?;* Alexander, *New Jim Crow;* Miller, *Halfway Home.*

26. Thompson and Eschmann, "Tweeting Abolition in an Age of Mass Incarceration."

27. Buchanan et al., "Black Lives Matter."

28. "Congress Must Divest the Billion Dollar Police Budget"; Garcetti, "Budget for the Fiscal Year 2019–2020"; Hamaji et al., *Freedom to Thrive;* Martin, "Look at the 'People's Budget.'"

29. McDowell and Fernandez, "'Disband, Disempower, and Disarm.'"

30. Miller, "NYC Is Sending Social Workers."

31. McDowell and Fernandez, "'Disband, Disempower, and Disarm'"; #8ToAbolition, "8 to Abolition—Why."

32. Breland, "How Twitter Vaulted 'Abolish ICE.'"

33. González-Ramírez, "Here's How the #AbolishICE Movement."

34. Ritchie et al., "Reparations Now Toolkit."

35. Coates, "Case for Reparations"; Rosario, "Unlikely Story Behind Japanese Americans.'"

36. Velshi, "Robin Rue Simmons"

37. Beam, "11 US Mayors Have Pledged."

38. "Episode 3: The House We Live." RACE—The Power of an Illusion.

39. Douglass, "The Meaning of the Fourth of July," 196.

40. Wise, "Juneteenth Is Now a Federal Holiday."

41. Viala-Gaudefroy and Lindaman, "Donald Trump's 'Chinese Virus.'"

42. Yam, "Viral Images Show People of Color."

43. Zhang et al., "Hate Crimes against Asian Americans."

44. Jerusalem Demsas, "History of Tensions—and Solidarity"; Ho and Clayton, "'Black and Asian Unity.'"

45. Ng et al., "Contesting the Model Minority."

46. Gadzo and Jubeh, "Sheikh Jarrah Is 'under a Siege.'"

47. "'If I Don't Steal [Your Home] Someone Else Will.'"

48. Erakat and Hill, "Black-Palestinian Transnational Solidarity."

49. Kelley, "From the River to the Sea to Every Mountain Top."

50. Chokshi, "Angela Davis Won an Award."

51. Kweli, *Vibrate Higher.*

52. Gutman, "Why Zionism and Antisemitism."

53. Hussain, "States Can't Control the Narrative."

54. Farzan, "Israeli-Palestinian Conflict."

55. A storm at least, by my standards, though when I finally made it to town I was laughed at for calling that two-foot "dusting" a storm. I was told that the snowbanks in January are typically human height, and when it really storms, people travel by snowmobile, as cars can't get anywhere, even after a night of plowing.

56. I had rented a 4x4, but forgot to turn on the all-wheel drive option, like a city genius.

57. My Airbnb host told me that losing power during the storm was a possibility, but not to worry as I had plenty of wood and flashlights.

58. Corasaniti, "Georgia G.O.P. Passes Major Law."

59. Lee, "Eddie Murphy and Spike Lee in Conversation."

60. "8:46—Dave Chapelle." Netflix Is a Joke.

61. hooks, *Yearning.*

62. Ghaffary, "How TikTok's Hate Speech Detection Tool"; Gebeily, "Instagram, Twitter Blame Glitches."

63. Chowdhury and Williams, "Introducing Twitter's First Algorithmic Bias Bounty Challenge."

APPENDIX

1. Hopkins and King, "Method of Automated Nonparametric Content Analysis."

2. Timmermans and Tavory, "Theory Construction in Qualitative Research."

3. Glaser and Strauss, *Discovery of Grounded Theory.*

4. Kroenke et al., "An Ultra-Brief Screening Scale"; Elo et al., "Validity of a Single-Item Measure."

5. Keels et al., "Psychological and Academic Costs."

References

#8ToAbolition. "8 to Abolition—Why." 2020. www.8toabolition.com/why.

"8:46—Dave Chapelle." Netflix Is a Joke. YouTube video, 27:20. June 12, 2020. www.youtube.com/watch?v=3tR6mKcBbT4&ab_channel=NetflixIsAJoke.

Abel, Elizabeth. *Signs of the Times: The Visual Politics of Jim Crow.* Berkeley: University of California Press, 2010.

Alba, Davey, Ella Koeze, and Jacob Silver. "What Happened When Trump Was Banned on Facebook and Twitter." *New York Times.* June 7, 2021. www.nytimes.com/interactive/2021/06/07/technology/trump-social-media-ban.html.

Albertson, Bethany L. "Dog-Whistle Politics: Multivocal Communication and Religious Appeals." *Political Behavior* 37, no. 1 (2015): 3–26.

Alegria, Margarita, Nicholas J. Carson, Marta Goncalves, and Kristen Keefe. "Disparities in Treatment for Substance Use Disorders and Co-occurring Disorders for Ethnic/Racial Minority Youth." *Journal of the American Academy of Child and Adolescent Psychiatry* 50, no. 1 (2011): 22–31.

Alexander, Michelle. *The New Jim Crow: Mass Incarceration in the Age of Color-blindness.* New York: New Press, 2012.

Amatulli, Jenna. "Trump Says He's 'The Least Racist Person in This Room' at Final Presidential Debate." HuffPost. October 22, 2020. www.huffpost.com /entry/trump-kristen-welker-least-racist_n_5f923e9dc5b686eaaa0fb460.

Anderson, Carol. *White Rage: The Unspoken Truth of Our Racial Divide.* New York: Bloomsbury Publishing USA, 2016.

Anderson, Monica, and Paul Hitlin. "Social Media Conversations about Race: How Social Media Users See, Share, and Discuss Race and the Rise of Hashtags like #BlackLivesMatter." Pew Research Center. August 15, 2016. www .pewresearch.org/internet/2016/08/15/the-hashtag-blacklivesmatter-emerges-social-activism-on-twitter/.

Andrew, Scottie. "Twitter Suspends Fake Accounts Pretending to Be Black Trump Supporters." CNN. October 14, 2020. www.cnn.com/2020/10/14 /tech/twitter-suspends-fake-black-trump-supporters-trnd/index.html.

Apple. "Use Face ID While Wearing a Mask with iPhone 12 and Later." Apple Support. Accessed June 25, 2022. https://support.apple.com/en-us/HT213062.

Auxier, Brooke, and Monica Anderson. "Social Media Use in 2021." Pew Research Center, *Science & Tech* (blog). April 7, 2021. www.pewresearch.org /internet/2021/04/07/social-media-use-in-2021/.

Bailey, Moya. *Misogynoir Transformed: Black Women's Digital Resistance.* New York: New York University Press, 2021.

———. "On Misogynoir: Citation, Erasure, and Plagiarism." *Feminist Media Studies* 18, no. 4 (2018): 762–68.

Baker, Mike. "'Murder Hornets' in the U.S.: The Rush to Stop the Asian Giant Hornet." *New York Times.* May 2, 2020. www.nytimes.com/2020/05/02/us /asian-giant-hornet-washington.html.

Banning, Marlia E. "The Limits of PC Discourse: Linking Language Use to Social Practice." *Pedagogy* 4, no. 2 (2004): 191–214.

Barberá, Pablo, John T. Jost, Jonathan Nagler, Joshua A. Tucker, and Richard Bonneau. "Tweeting from Left to Right: Is Online Political Communication More Than an Echo Chamber?" *Psychological Science* 26, no. 10 (October 2015): 1531–42.

Bartlett, Jamie, Jeremy Reffin, Noelle Rumball, and Sarah Williamson. *Antisocial Media.* London: Demos, 2014. http://cilvektiesibas.org.lv/site/record /docs/2014/03/19/DEMOS_Anti-social_Media.pdf.

Beam, Adam. "11 US Mayors Have Pledged to Pay Reparations for Slavery to a Small Group of Black Residents in Their Cities." *Chicago Tribune.* June 18, 2021. www.chicagotribune.com/nation-world/ct-aud-nw-us-mayors-reparations-slavery-payments-20210618-zoi3rvfkabcbpfqiawfzaohfmu-story.html.

Benjamin, Ruha. *Race after Technology: Abolitionist Tools for the New Jim Code.* Cambridge: Polity, 2019.

Berenson, Tessa. "David Duke Says He and Donald Trump Have the Same Message." *Time.* September 30, 2016. https://time.com/4514350/david-duke-donald-trump-senate-louisiana/.

"Bill Clinton the First Black President? Bust a Move, Jokes Barack Obama." *New York Daily News.* January 22, 2008. www.nydailynews.com/news/politics/bill-clinton-black-president-bust-move-jokes-barack-obama-article-1.342196.

Binkley, Collin. "Harvard Pledges $100 Million to Atone for Role in Slavery." NBC Boston. April 26, 2022. www.nbcboston.com/news/local/harvard-pledges-100-million-to-atone-for-role-in-slavery/2703907/.

Binoya, Addy. "Racism in Porn Industry: Black Performers Reveal They're Paid and Hired Less." International Business Times. June 12, 2020. www.ibtimes.com/racism-porn-industry-black-performers-reveal-theyre-paid-hired-less-2992749.

Black Lives Matter. "Black Lives Matter: About." Accessed June 26, 2022. https://blacklivesmatter.com/about/.

Black, Lora L., Rhonda Johnson, and Lisa VanHoose. "The Relationship between Perceived Racism/Discrimination and Health among Black American Women: A Review of the Literature from 2003 to 2013." *Journal of Racial and Ethnic Health Disparities* 2, no. 1 (2015): 11–20.

Black, Marc. "Fanon and DuBoisian Double Consciousness." *Human Architecture: Journal of the Sociology of Self-Knowledge* 5, no. 3 (2007): 393–404.

Blake, Aaron. "No, Biden's New Border Move Isn't Like Trump's 'Kids in Cages.'" *Washington Post.* February 23, 2021. www.washingtonpost.com/politics/2021/02/23/no-bidens-new-border-move-isnt-like-trumps-kids-cages-not-hardly/.

Blume, Arthur W., Laura V. Lovato, Bryan N. Thyken, and Natasha Denny. "The Relationship of Microaggressions with Alcohol Use and Anxiety among Ethnic Minority College Students in a Historically White

Institution." *Cultural Diversity and Ethnic Minority Psychology* 18, no. 1 (January 2012): 45–54.

Bobo, Lawrence. "Racism in Trump's America: Reflections on Culture, Sociology, and the 2016 US Presidential Election." *British Journal of Sociology* 68, no. S1 (2017): S85–S104. https://doi.org/10.1111/1468-4446.12324.

Bobo, Lawrence, James R. Kluegel, and Ryan A. Smith. "Laissez-Faire Racism: The Crystallization of a Kinder, Gentler, Anti-Black Ideology." *Racial Attitudes in the 1990s: Continuity and Change,* edited by Steven A. Tuch and Jack K. Martin. Westport, CT: Praeger.

Bond, Robert M., Christopher J. Fariss, Jason J. Jones, Adam D. I. Kramer, Cameron Marlow, Jaime E. Settle, and James H. Fowler. "A 61-Million-Person Experiment in Social Influence and Political Mobilization." *Nature* 489 (2012): 295–98.

Bond, Shannon. "Elon Musk Says He'll Reverse Donald Trump Twitter Ban." NPR. May 10, 2022. www.npr.org/2022/05/10/1097942860/elon-musk-reverse-donald-trump-twitter-ban.

Bonilla-Silva, Eduardo. "The Linguistics of Color Blind Racism: How to Talk Nasty about Blacks without Sounding 'Racist.'" *Critical Sociology* 28, no. 1–2 (2002): 41–64.

———. *Racism without Racists: Color-Blind Racism and the Persistence of Racial Inequality in America.* Lanham, MD: Rowman & Littlefield, 2017.

———. *Racism without Racists: Color-Blind Racism and the Persistence of Racial Inequality in America.* 6th ed. Lanham, MD: Rowman & Littlefield, 2021.

———. "The Structure of Racism in Color-Blind, 'Post-Racial' America." *American Behavioral Scientist* 59, no. 11 (2015): 1358–76.

Boulianne, Shelley. "Social Media Use and Participation: A Meta-Analysis of Current Research." *Information, Communication & Society* 18, no. 5 (2015): 524–38.

Boyd, Danah. *It's Complicated: The Social Lives of Networked Teens.* New Haven, CT: Yale University Press, 2014.

Brandom, Russell. "Why Platforms Had to Cut off Trump and Parler." The Verge. January 11, 2021. www.theverge.com/22224860/parler-trump-deplatformed-capitol-raid-moderation-censorship-facebook-amazon-twitter.

Breland, Ali. "How Twitter Vaulted 'Abolish ICE' into the Mainstream." The Hill. July 29, 2018. thehill.com/policy/technology/399303-how-twitter-vaulted-abolish-ice-into-the-mainstream.

Broadwater, Luke, and Michael S. Schmidt. "Trump Urged Armed Supporters to Capitol, White House Aide Testifies." *New York Times.* June 28, 2022. www .nytimes.com/2022/06/28/us/politics/trump-meadows-jan-6-surprise-hearing .html.

Brock, André. "Race Matters: African Americans on the Web following Hurricane Katrina." In *Proceedings of Cultural Attitudes towards Communication and Technology,* edited by F. Sudweeks, H. Hrachovec, and C. Ess, 91–105. School of Information Technology, Murdoch University, 2008.

———. "Critical Technocultural Discourse Analysis." *New Media and Society* 20, no. 3 (2018): 1012–30.

Brock, André, Jr. *Distributed Blackness: African American Cybercultures.* New York: New York University Press, 2020.

Brondolo, Elizabeth, Erica E. Love, Melissa Pencille, Antoinette Schoenthaler, and Gbenga Ogedegbe. "Racism and Hypertension: A Review of the Empirical Evidence and Implications for Clinical Practice." *American Journal of Hypertension* 24, no. 5 (2011): 518–29.

Brown, Melissa, Rashawn Ray, Ed Summers, and Neil Fraistat. "#SayHerName: A Case Study of Intersectional Social Media Activism." *Ethnic and Racial Studies* 40, no. 11 (2017): 1831–46.

Buchanan, Larry, Quoctrung Bui, and Jugal K. Patel. "Black Lives Matter May Be the Largest Movement in U.S. History." *New York Times.* July 3, 2020. www.nytimes.com/interactive/2020/07/03/us/george-floyd-protests-crowd-size.html.

Bush, Harold K. "A Brief History of PC, with Annotated Bibliography." *American Studies International* 33, no. 1 (1995): 42–64.

Cabrera, Nolan L. "'But I'm Oppressed Too': White Male College Students Framing Racial Emotions as Facts and Recreating Racism." *International Journal of Qualitative Studies in Education* 27, no. 6 (2014): 768–84.

Carlisle, Kate. "Can We Talk? On College Campuses—Including Mayflower Hill—Free Speech Collides with Political Correctness." *Colby Magazine* 104, no. 3 (2016): 14.

Carter, Dan T. *From George Wallace to Newt Gingrich: Race in the Conservative Counterrevolution, 1963–1994.* Baton Rouge: Louisiana State University Press, 1996.

Case, Andrew D., and Carla D. Hunter. "Counterspaces: A Unit of Analysis for Understanding the Role of Settings in Marginalized Individuals' Adaptive

Responses to Oppression." *American Journal of Community Psychology* 50, no. 1–2 (September 2012): 257–70.

Chavez, Leo. *The Latino Threat: Constructing Immigrants, Citizens, and the Nation.* 2nd ed. Stanford, CA: Stanford University Press, 2013.

Chokshi, Niraj. "Angela Davis Won an Award. It Was Revoked. Now It's Been Reinstated." January 25, 2019. www.nytimes.com/2019/01/25/us/angela-davis-israel.html?.?mc=aud_dev&ad-keywords=auddevgate&gclid=CjwKC Ajw2P-KBhByEiwADBYWCqMVDVANF9Zxc9GTmCQ2O84qp0_SuT3iK-MiAsBQzaR4dupJB14LinBoC7SgQAvD_BwE&gclsrc=aw.ds.

Chowdhury, Rumman, and Jutta Williams. "Introducing Twitter's First Algorithmic Bias Bounty Challenge." Twitter Engineering Blog: Insights. July 30, 2021. https://blog.twitter.com/engineering/en_us/topics/insights/2021/algorithmic-bias-bounty-challenge.

Christopherson, Kimberly M. "The Positive and Negative Implications of Anonymity in Internet Social Interactions: 'On the Internet, Nobody Knows You're a Dog.'" *Computers in Human Behavior* 23, no. 6 (2007): 3038–56.

Coates, Ta-Nehisi. "The Case for Reparations." *Atlantic,* June 2014. www.theatlantic.com/magazine/archive/2014/06/the-case-for-reparations/361631/.

Cogburn, Courtney D., Jeremy Bailenson, Elise Ogle, Tobin Asher, and Teff Nichols. *1000 Cut Journey.* Virtual, Augmented, and Mixed Reality. ACM SIGGRAPH. 2018. ACM Digital Library. https://dl.acm.org/doi/10.1145/3226552.3226575.

Cohen, Cathy. "Reimagining Political Participation in the Digital Age." Youth and Participatory Politics Research Network. 2016. https://ypp.dmlcentral.net/publications/309.

Cohen, Cathy J. "Deviance as Resistance: A New Research Agenda for the Study of Black Politics." *Du Bois Review: Social Science Research on Race* 1, no. 1 (March 2004): 27–45.

Cohen, Cathy J., and Joseph Kahne. "Participatory Politics: New Media and Youth Political Action." Youth and Participatory Politics Research Network. 2012. https://ypp.dmlcentral.net/sites/default/files/publications/Participatory_Politics_New_Media_and_Youth_Political_Action.2012.pdf.

Colleoni, Elanor, Alessandro Rozza, and Adam Arvidsson. "Echo Chamber or Public Sphere? Predicting Political Orientation and Measuring Political

Homophily in Twitter Using Big Data: Political Homophily on Twitter." *Journal of Communication* 64, no. 2 (April 2014): 317–32.

Combahee River Collective. "The Combahee River Collective Statement." 1977. Posted online on November 16, 2012. BlackPast.org. https://www.blackpast.org/african-american-history/combahee-river-collective-statement-1977/.

"Congress Must Divest the Billion Dollar Police Budget and Invest in Public Education." The Center for Popular Democracy. June 10, 2020. https://www.populardemocracy.org/news-and-publications/congress-must-divest-billion-dollar-police-budget-and-invest-public-education.

Conroy, Meredith, Jessica T. Feezell, and Mario Guerrero. "Facebook and Political Engagement: A Study of Online Political Group Membership and Offline Political Engagement." *Computers in Human Behavior* 28, no. 5 (2012): 1535–46.

Cook, Gina, and Jackie Bensen. "Racist Photo Sparks Outrage among GW University Students." NBC4 Washington. February 2, 2018. http://www.nbcwashington.com/news/local/George-Washington-Univeristy-Investigating-Racist-Photo-Posted-on-Snapchat-472247113.html.

Corasaniti, Nick. "Georgia G.O.P. Passes Major Law to Limit Voting, Part of Nationwide Push." *New York Times.* March 25, 2021. https://www.nytimes.com/2021/03/25/us/politics/georgia-voting-law-republicans.html.

Correll, Joshua, Bernadette Park, Charles M. Judd, Bernd Wittenbrink, Melody S. Sadler, and Tracie Keesee. "Across the Thin Blue Line: Police Officers and Racial Bias in the Decision to Shoot." *Journal of Personality and Social Psychology* 92, no. 6 (2007): 1006–23.

Correll, Joshua, Sean M. Hudson, Steffanie Guillermo, and Debbie S. Ma. "The Police Officer's Dilemma: A Decade of Research on Racial Bias in the Decision to Shoot." *Social and Personality Psychology Compass* 8, no. 5 (2014): 201–13.

Cottom, Tressie McMillan. "Where Platform Capitalism and Racial Capitalism Meet: The Sociology of Race and Racism in the Digital Society." *Sociology of Race and Ethnicity* 6, no. 4 (2020): 441–49. https://doi.org/10.1177/2332649220949473.

———. "Black Twitter Is Not a Place. It's a Practice." *New York Times.* May 3, 2022. https://www.nytimes.com/2022/05/03/opinion/the-real-twitter-is-not-for-sale.html.

Crenshaw, Kimberlé. "Mapping the Margins: Intersectionality, Identity Politics, and Violence against Women of Color." *Stanford Law Review* 43, no. 6 (1991): 1241–1300.

Crimmins, Eileen M., and Yasuhiko Saito. "Trends in Healthy Life Expectancy in the United States, 1970–1990: Gender, Racial, and Educational Differences." *Social Science and Medicine* 52, no. 11 (2001): 1629–41.

"Cuomo to Trump Supporter: The Facts Are Not Your Friend." CNN video, 1:04. Cuomo Prime Time. 2019. https://www.cnn.com/videos/politics/2019/07/18/cuomo-great-debate-trump-racist-remarks-democratic-congresswomen-cpt-vpx.cnn.

"Cuomo: What Would You Do If Trump Said "I Am a Racist"?" CNN video, 2:48. 2019. https://www.cnn.com/videos/politics/2019/07/17/kris-kobach-asked-would-he-support-admittedly-racist-president-trump-cpt-sot-vpx.cnn.

Daniels, Jessie. "Cloaked Websites: Propaganda, Cyber-Racism and Epistemology in the Digital Era." *New Media & Society* 11, no. 5 (2009): 659–83.

———. *Cyber Racism: White Supremacy Online and the New Attack on Civil Rights.* Lanham, MD: Rowman & Littlefield, 2009.

———. "Race and Racism in Internet Studies: A Review and Critique." *New Media & Society* 15, no. 5 (2013): 695–719.

———. "The Algorithmic Rise of the 'Alt-Right.'" *Contexts* 17, no. 1 (February 2018): 60–65. https://doi.org/10.1177/1536504218766547.

Davis, Angela. *Are Prisons Obsolete?* New York: Seven Stories Press, 2003.

Dazey, Margot. "Rethinking Respectability Politics." *British Journal of Sociology* 72, no. 3 (2021): 580–93.

Delgado, Richard. "Legal Realism and the Controversy over Campus Speech Codes." *Case Western Reserve Law Review* 69 (2018): 275–98.

Delgado, Richard, and Jean Stefancic. *Critical Race Theory: An Introduction.* New York: New York University Press, 2017.

———. *Critical Race Theory: The Cutting Edge.* Philadelphia: Temple University Press, 2000.

Demsas, Jerusalem. "The History of Tensions—and Solidarity—between Black and Asian American Communities, Explained." Vox. March 16, 2021. www.vox.com/22321234/black-asian-american-tensions-solidarity-history.

DeWitt, Larry. "The Decision to Exclude Agricultural and Domestic Workers from the 1935 Social Security Act." *Social Security Bulletin* 70, no. 4 (2010). Social Security: Office of Retirement and Disability Policy. www.ssa.gov /policy/docs/ssb/v70n4/v70n4p49.html.

Depenbrock, Julie. "Federal Judge Finds Racism behind Arizona Law Banning Ethnic Studies." NPR. August 22, 2017. https://www.npr.org/sections/ed /2017/08/22/545402866/federal-judge-finds-racism-behind-arizona-law-banning-ethnic-studies.

DiAngelo, Robin. "White Fragility." *International Journal of Critical Pedagogy* 3, no. 3 (2011): 54–70. http://libjournal.uncg.edu/index.php/ijcp/article/view /249%3E.

"Do You Think Bill Clinton Was Our First Black President? [CNN]." YouTube video, 2:58. January 22, 2008. www.youtube.com/watch?v= zARV48q8Cao.

Domonoske, Camila. "Trump Fails to Condemn KKK on Television, Turns to Twitter to Clarify." NPR. February 28, 2016. www.npr.org/sections /thetwo-way/2016/02/28/468455028/trump-wont-condemn-kkk-says-he-knows-nothing-about-white-supremacists.

Douglass, Frederick. "The Meaning of the Fourth of July for the Negro." A speech given at Rochester, New York, July 5, 1852. Available at https:// masshumanities.org/files/programs/douglass/speech_complete.pdf.

Downs, Kenya. "FBI Warned of White Supremacists in Law Enforcement 10 Years Ago. Has Anything Changed?" PBS NewsHour. October 21, 2016. www .pbs.org/newshour/nation/fbi-white-supremacists-in-law-enforcement.

Du Bois, W. E. B. *The Souls of Black Folk*. Barnes & Noble Classics. New York: Barnes & Noble Books, 2003.

Duster, Troy. "The Diversity Project: Final Report." University of California, Berkeley, Institute for the Study of Social Change, 1991.

Dwyer, Colin. "Donald Trump: 'I Could . . . Shoot Somebody, And I Wouldn't Lose Any Voters.'" NPR. January 23, 2016. www.npr.org/sections/thetwo-way/2016/01/23/464129029/donald-trump-i-could-shoot-somebody-and-i-wouldnt-lose-any-voters.

Eagly, Alice H., and Amanda B. Diekman. "What Is the Problem? Prejudice as an Attitude-in-Context." In *On the Nature of Prejudice: Fifty Years after Allport,* edited by John F. Dovidio, Peter Glick, and Laurie A. Rudman, 19–35. Malden, MA: Blackwell, 2005.

Eisgruber, Christopher L. "Op-ed by Princeton President Eisgruber on Free Speech and Inclusivity: Why Mutual Respect Makes Free Speech Better" (blog). Princeton School of Public and International Affairs. July 20, 2020. https://spia.princeton.edu/blogs/op-ed-princeton-president-eisgruber-free-speech-and-inclusivity-why-mutual-respect-makes-free.

Ellis, Ralph. "Former Kentucky Police Official Allegedly Sent Racist Messages to Recruit." CNN. January 22, 2018. www.cnn.com/2018/01/22/us/kentucky-officer-urged-recruit-to-shoot-blacks/index.html.

Elo, Anna-Liisa, Anneli Leppänen, and Antti Jahkola. "Validity of a Single-Item Measure of Stress Symptoms." *Scandinavian Journal of Work, Environment & Health* 29, no. 6 (2003): 444–51.

"Episode 3: The House We Live." RACE—The Power of an Illusion. The online companion to California Newsreel's 3-part documentary about race in society, science, and history. PBS. 2003. www.pbs.org/race/000_General/000_00-Home.htm.

Erakat, Noura, and Marc Lamont Hill. "Black-Palestinian Transnational Solidarity: Renewals, Returns, and Practice." *Journal of Palestine Studies* 48, no. 4 (2019): 7–16.

Eschmann, Rob. "Digital Resistance: How Online Communication Facilitates Responses to Racial Microaggressions." *Sociology of Race and Ethnicity* 7, no. 2 (2021): 264–77.

———. "Unmasking Racism: Students of Color and Expressions of Racism in Online Spaces." *Social Problems* 67, no. 3 (2020): 418–36.

Eschmann, Rob, Jacob Groshek, Rachel Chanderdatt, Khea Chang, and Maysa Whyte. "Making a Microaggression: Using Big Data and Qualitative Analysis to Map the Reproduction and Disruption of Microaggressions through Social Media." *Social Media + Society* 6, no. 4 (2020). https://doi.org/10.1177/2056305120975716.

Eschmann, Rob, Jacob Groshek, Senhao Li, Noor Turaif, and Julian Thompson. "Bigger than Sports: Identity Politics, Colin Kaepernick, and Concession Making in #BoycottNike." *Computers in Human Behavior* 114 (2021): 106583. https://doi.org/10.1016/j.chb.2020.106583.

Eschmann, Rob, Julian G. Thompson, and Noor Toraif. "Tweeting toward Transformation: Prison Abolition and Criminal Justice Reform in 140 Characters." *Sociological Inquiry*, 2022. https://doi.org/10.1111/soin.12503

Eschmann, Rob, Ryan W. Gryder, Gerri Connaught, Xiang Zhao, Sae-Mi Jeon, and Ernest Gonzales. "Context Matters: Differential Effects of Discrimination by Environmental Context on Depressive Symptoms Among College Students of Color." *Clinical Social Work Journal*. 50 (2021): 242–55.

Eschmann, Robert Daniel. "How the Internet Shapes Racial Discourse: Students of Color, Racism, and Resistance in Online Spaces." PhD diss, University of Chicago, 2017.

Eustachewich, Lia. "White Male New Hampshire Professor Allegedly Posed as Woman of Color on Twitter." *New York Post* (blog). October 6, 2020. https://nypost.com/2020/10/06/white-male-new-hampshire-professor-allegedly-posed-as-woman-of-color/.

Evans, Louwanda, and Wendy Leo Moore. "Impossible Burdens: White Institutions, Emotional Labor, and Micro-Resistance." *Social Problems* 62, no. 3 (August 2015): 439–54. https://doi.org/10.1093/socpro/spv009.

Everett, Ronald S., and Roger A. Wojtkiewicz. "Difference, Disparity, and Race/Ethnic Bias in Federal Sentencing." *Journal of Quantitative Criminology* 18, no. 2 (2002): 189–211.

Falcón, Sylvanna M. "Mestiza Double Consciousness: The Voices of Afro-Peruvian Women on Gendered Racism." *Gender and Society* 22, no. 5 (2008): 660–80.

Farzan, Antonia Noori. "Israeli-Palestinian Conflict Plays out on Social Media as Activists Raise Concerns about Instagram, Twitter Interference." *Washington Post*. May 12, 2021. www.washingtonpost.com/world/2021/05/12/israeli-palestinian-conflict-social-media/.

Faust, Drew Gilpin. "Harvard University 2017 Commencement Speech." Harvard University. May 25, 2017. www.harvard.edu/president/speech/2017/2017-commencement-speech.

FBI Counterterrorism Division. "(U) White Supremacist Infiltration of Law Enforcement," October 17, 2006. https://s3.documentcloud.org/documents/3439212/FBI-White-Supremacist-Infiltration-of-Law.pdf.

Fernando, Sumam. "Racism as a Cause of Depression." *International Journal of Social Psychiatry* 30, no. 1–2 (1984): 41–49.

Folayan, Sabaah, and Damon Davis, dir. *Whose Streets?* New York: Magnolia Pictures, 2017.

Frankenberg, Erica, Jongyeon Ee, Jennifer B. Ayscue, and Gary Orfield. "Harming Our Common Future: America's Segregated Schools 65 Years after Brown." The Civil Rights Project, UCLA, and Center for Education and Civil Rights. 2019. www.civilrightsproject.ucla.edu/research /k-12-education/integration-and-diversity/harming-our-common-future-americas-segregated-schools-65-years-after-brown/Brown-65–050919v4-final .pdf.

Fryer, Roland G. "An Empirical Analysis of Racial Differences in Police Use of Force." *Journal of Political Economy* 127, no. 3 (October 30, 2018): 1210–61.

Gadzo, Mersiha, and Dareen Jubeh. "Sheikh Jarrah Is 'under a Siege,' Palestinian Residents Say." Al Jazeera. May 21, 2021. www.aljazeera.com/news/2021 /5/21/we-dont-sleep-at-night-palestinians-in-sheikh-jarrah.

Garcetti, Eric. "Budget for the Fiscal Year 2019–2020." City of Los Angeles. 2019. Retrieved from http://cao.lacity.org/budget19-20/2019-20Proposed_Budget .pdf.

Gates, Henry Louis, and Cornel West. *The Future of the Race.* New York: Vintage, 1997.

Gebeily, Maya. "Instagram, Twitter Blame Glitches for Deleting Palestinian Posts." Reuters. May 10, 2021. www.reuters.com/article/israel-palestinians-socialmedia-idUSL8N2MU624.

Ghaffary, Shirin. "How TikTok's Hate Speech Detection Tool Set off a Debate about Racial Bias on the App." Vox. July 7, 2021. www.vox.com/recode/2021 /7/7/22566017/tiktok-black-creators-ziggi-tyler-debate-about-black-lives-matter-racial-bias-social-media.

Gladwell, Malcolm. "Small Change: Why the Revolution Will Not Be Tweeted." *New Yorker.* September 27, 2010. www.newyorker.com/magazine/2010 /10/04/small-change-malcolm-gladwell.

Glaser, Barney G., and Anselm L. Strauss. *The Discovery of Grounded Theory: Strategies for Qualitative Research.* Chicago: Aldine, 1967.

Goldberg, Suzanne B. "Speech, Protest and the Rules of University Conduct." Columbia University Life. February 11, 2017. https://universitylife.columbia .edu/Speech-Protest-Rules.

González-Ramírez, Andrea. "Here's How the #AbolishICE Movement Really Got Started." Refinery 29. July 30, 2018. www.refinery29.com/en-us/2018/07 /205854/abolish-ice-origins-twitter-undocumented-immigrants.

Goudie, Chuck, Barb Markoff, Christine Tressel, Ross Weidner, and Jonathan Fagg. "Chicago Has Authorized Nearly $67M in Police Misconduct Settlement Payments So Far This Year." ABC7 Chicago. December 13, 2021. https://abc7chicago.com/chicago-police-department-misconduct-payout /11336008/.

Graham, Jessica R., Lindsey M. West, Jennifer Martinez, and Lizabeth Roemer. "The Mediating Role of Internalized Racism in the Relationship between Racist Experiences and Anxiety Symptoms in a Black American Sample." *Cultural Diversity and Ethnic Minority Psychology* 22, no. 3 (2016): 369–76.

Gray, Kishonna L. "Intersecting Oppressions and Online Communities: Examining the Experiences of Women of Color in Xbox Live." *Information, Communication & Society* 15, no. 3 (April 2012): 411–28.

———. *Intersectional Tech: Black Users in Digital Gaming.* Baton Rouge: Louisana State University Press, 2020.

Gregory, Anne, Russell J. Skiba, and Pedro A. Noguera. "The Achievement Gap and the Discipline Gap: Two Sides of the Same Coin?" *Educational Researcher* 39, no. 1 (2010): 59–68.

Grimm, Pamela. "Social Desirability Bias." In *Wiley International Encyclopedia of Marketing,* edited by J. Sheth and N. Malhotra, 2010. Hoboken, NJ: John Wiley. https://onlinelibrary.wiley.comdoi/full/10.1002/9781444316568.

Groshek, Jacob, and Chelsea Cutino. "Meaner on Mobile: Incivility and Impoliteness in Communicating Online." In *Proceedings of the 7th 2016 International Conference on Social Media and Society* (ACM, 2016).

Groshek, Jacob, and Edson Tandoc. "The Affordance Effect: Gatekeeping and (Non)Reciprocal Journalism on Twitter." *Computers in Human Behavior* 66 (January 2017): 201–10.

Gutman, Abraham. "Why Zionism and Antisemitism Are Each Other's Best Recruiting Tools." NBC News. May 27, 2021. www.nbcnews.com/think /opinion/how-jews-can-support-palestinian-rights-condemn-antisemitism-ncna1268680.

Hall, Jacquelyn Dowd. "The Long Civil Rights Movement and the Political Uses of the Past." *Journal of American History* 91, no. 4 (2005): 1233–63.

Hall, Madison, Skye Gould, Rebecca Harrington, Jacob Shamsian, Azmi Haroun, Taylor Ardrey, and Erin Snodgrass. "At Least 865 People Have Been Charged in the Capitol Insurrection So Far. This Searchable Table Shows Them All." Insider. June 13, 2022. www.insider.com/all-the-us-capitol-pro-trump-riot-arrests-charges-names-2021–1.

Haltiwanger, John. "Trump's History of Support from White Supremacist, Far Right Groups." Business Insider. September 30, 2020. www.businessinsider.com/trumps-history-of-support-from-white-supremacist-far-right-groups-2020-9.

Hamaji, Kate, Kumar Rao, Marbre Stahly-Butts, Janaé Bonsu, Charlene A. Carruthers, Roselyn Berry, and Denzel McCampbell. *Freedom to Thrive: Reimagining Safety & Security in Our Communities.* Center for Popular Democracy, 2017. www.populardemocracy.org/news/publications/freedom-thrive-reimagining-safety-security-our-communities.

Hawkins, Stephen, Daniel Yudkin, Míriam Juan-Torres, and Tim Dixon. "Hidden Tribes: A Study of America's Polarized Landscape," 2018. A report from More in Common. https://hiddentribes.us/media/qfpekz4g/hidden_tribes_report.pdf.

Heim, Joe. "Recounting a Day of Rage, Hate, Violence and Death: How a Rally of White Nationalists and Supremacists at the University of Virginia Turned into a 'Tragic, Tragic Weekend.'" *Washington Post.* August 14, 2017. www.washingtonpost.com/graphics/2017/local/charlottesville-timeline/.

Hermann, Peter, Melissa J. Lang, and Clarence Williams. "Pro-Trump Rally Descends into Chaos as Proud Boys Roam D.C. Looking to Fight." *Washington Post.* December 13, 2020. www.washingtonpost.com/local/public-safety/proud-boys-protest-stabbing-arrest/2020/12/13/98c0f740-3d3f-11eb-8db8-395dedaaa036_story.html.

Herreria, Carla. "Woman Calls Police on Black Family for BBQing at a Lake in Oakland" HuffPost. May 11, 2018. www.huffingtonpost.com/entry/woman-calls-police-oakland-barbecue_us_5af50125e4b00d7e4c18f741.

Herron, Arika. "Bill That Made Tempers Rise Has Worrying Feel to Some." *Indianapolis Star.* March 1, 2021.

Herron, Rachael. "I Used to be a 911 Dispatcher. I Had to Respond to Racist Calls Every Day." Vox. October 31, 2018. www.vox.com/first-person/2018/5

/30/17406092/racial-profiling-911-bbq-becky-living-while-black-babysitting-while-black.

Hill, Marc Lamont. "'Thank You, Black Twitter': State Violence, Digital Counterpublics, and Pedagogies of Resistance." *Urban Education* 53, no. 2 (February 2018): 286–302. https://doi.org/10.1177/0042085917747124.

Ho, Vivian, and Abené Clayton. "'Black and Asian Unity': Attacks on Elders Spark Reckoning with Racism's Roots." *Guardian* (US edition). February 21, 2021. www.theguardian.com/us-news/2021/feb/21/black-and-asian-unity-attacks-on-elders-spark-reckoning-with-racisms-roots.

Hoggard, Lori S., Shawn C. T. Jones, and Robert M. Sellers. "Racial Cues and Racial Identity Implications for How African Americans Experience and Respond to Racial Discrimination." *Journal of Black Psychology* 43, no. 4 (2016): 409–32. https://doi.org/10.1177/0095798416651033.

hooks, bell. *Yearning: Race, Gender, and Cultural Politics*. Boston: South End Press, 1990.

Hopkins, Daniel J., and Gary King. "A Method of Automated Nonparametric Content Analysis for Social Science." *American Journal of Political Science* 54, no. 1 (January 2010): 229–47.

Hughes, Diane L., James Rodriguez, Emilie P. Smith, Deborah J. Johnson, Howard C. Stevenson, and Paul Spicer. "Parents' Ethnic-Racial Socialization Practices: A Review of Research and Directions for Future Study." *Developmental Psychology* 42, no. 5 (2006): 747–70.

Hughey, Matthew W., and Jessie Daniels. "Racist Comments at Online News Sites: A Methodological Dilemma for Discourse Analysis." *Media, Culture and Society* 35, no. 3 (2013): 332–47.

Hussain, Murtaza. "States Can't Control the Narrative on Israel-Palestine Anymore." *Intercept*. May 12, 2021. https://theintercept.com/2021/05/12/israel-palestine-jerusalem-social-media/.

"'If I Don't Steal [Your Home] Someone Else Will' Israeli Settler Justifies Forcible Takeover." YouTube video, 1:05. 2021. www.youtube.com/watch?v=KNqozQ8uaV8.

Islam, Inaash. "Muslim American Double Consciousness." *Du Bois Review: Social Science Research on Race* 17, no. 2 (2020): 429–48.

Jackson, Sarah J., Moya Bailey, and Brooke Foucault Welles. *#HashtagActivism: Networks of Race and Gender Justice*. Cambridge, MA: MIT Press, 2020.

Jacobo, Julia. "A Visual Timeline on How the Attack on Capitol Hill Unfolded." ABC News. January 10, 2021. https://abcnews.go.com/US/visual-timeline-attack-capitol-hill-unfolded/story?id=75112066.

Jaschick, Scott. "Incidents Roil Campuses." February 22, 2017. Inside Higher Ed. www.insidehighered.com/news/2017/02/22/racial-incidents-upset-students-several-campuses.

———. "The Chicago Letter and Its Aftermath." Inside Higher Ed. August 29, 2016. www.insidehighered.com/news/2016/08/29/u-chicago-letter-new-students-safe-spaces-sets-intense-debate.

Jeffrey, Adam, and Riya Bhattacharjee. "Photos Show Violent Clashes as Trump Supporters Storm the U.S. Capitol." CNBC. January 6, 2021. www.cnbc.com/2021/01/06/trump-supporters-rally-in-washington-to-protest-election-results.html.

Jones, Athena, and Tom Kludt. "Barack Obama to Black Voters: Trump Will Undo My Legacy." CNNPolitics. November 2, 2016. https://www.cnn.com/2016/11/02/politics/barack-obama-black-voters-donald-trump/index.html.

Jones, Feminista. *Reclaiming Our Space: How Black Feminists Are Changing the World from the Tweets to the Streets*. Boston: Beacon Press, 2019.

Jones, Ja'han. "Texas Senate Joins GOP Voter Suppression Push, Passes Restrictive SB 7 Bill." HuffPost. April 1, 2021. https://www.huffpost.com/entry/texas-senate-joins-voter-suppression-push-passes-bill-restricting-voter-access_n_6066076cc5b67785b7776971.

Jones, Leslie Kay. "#BlackLivesMatter: An Analysis of the Movement as Social Drama." *Humanity and Society* 44, no. 1 (2020): 92–110. https://doi.org/10.1177/0160597619832049.

Joravsky, Ben. "Did the NBA Blacklist Former Chicago Bulls Player Craig Hodges Because of His Political Beliefs?" Chicago Reader. December 14, 2016. http://chicagoreader.com/news-politics/did-the-nba-blacklist-former-chicago-bulls-player-craig-hodges-because-of-his-political-beliefs/.

Jordan, Michael. "Michael Jordan: 'I Can No Longer Stay Silent.'" Andscape. July 25, 2016. https://andscape.com/features/michael-jordan-i-can-no-longer-stay-silent/.

Jurgensen, Nathan. "Digital Dualism versus Augmented Reality." *Cyborgology* (blog). February 24, 2011. https://thesocietypages.org/cyborgology/2011/02/24/digital-dualism-versus-augmented-reality/.

Kafai, Yasmin B., Melissa S. Cook, and Deborah A. Fields. "'Blacks Deserve Bodies Too!': Design and Discussion About Diversity and Race in a Tween Virtual World." *Games and Culture* 5, no. 1 (2010): 43–63.

Kamenetz, Anya, Kayla Lattimore, and Julie Depenbrock. "Harvard Rescinds Admission Of 10 Students over Obscene Facebook Messages." NPR. 2017. http://www.npr.org/sections/ed/2017/06/06/531591202/harvard-rescinds-admission-of-10-students-over-obscene-facebook-messages.

Kang, Jerry. "Cyber-Race." *Harvard Law Review* 113 (2000): 1130–1208.

Kawakami, Kerry, Elizabeth Dunn, Francine Karmali, and John F. Dovidio. "Mispredicting Affective and Behavioral Responses to Racism." *Science* 323 (2009): 276–78.

Keels, Micere, Myles Durkee, and Elan Hope. "The Psychological and Academic Costs of School-Based Racial and Ethnic Microaggressions." *American Educational Research Journal* 54, no. 6 (December 2017): 1316–44.

Kelley, Robin D. G. "From the River to the Sea to Every Mountain Top: Solidarity as Worldmaking." *Journal of Palestine Studies* 48, no. 4 (2019): 69–91.

Kelly, Caroline. "Trump Bars 'Propaganda' Training Sessions on Race in Latest Overture to His Base." CNN. September 5, 2020. www.cnn.com/2020/09/04/politics/trump-administration-memo-race-training-ban/index.html.

Kendi, Ibram X. *How to Be an Antiracist*. New York: Random House, 2019.

———. *Stamped from the Beginning: The Definitive History of Racist Ideas in America*. New York: Nation Books, 2016.

Kennedy, Randall L. "Who Can Say 'Nigger'? And Other Considerations." *Journal of Blacks in Higher Education* 26 (Winter 1999–2000): 86–96.

Kil, Sang Hea. "Fearing Yellow, Imagining White: Media Analysis of the Chinese Exclusion Act of 1882." *Social Identities* 18, no. 6 (November 2012): 663–77.

Klein, Rick. "Trump Said 'Blame on Both Sides' in Charlottesville, Now the Anniversary Puts Him on the Spot." ABC News. August 12, 2018. https://abcnews.go.com/Politics/trump-blame-sides-charlottesville-now-anniversary-puts-spot/story?id=57141612.

Kramer, Rory, and Brianna Remster. "Stop, Frisk, and Assault? Racial Disparities in Police Use of Force during Investigatory Stops." *Law and Society Review* 52, no. 4 (2018): 960–93.

Kristofferson, Kirk, Katherine White, and John Peloza. "The Nature of Slacktivism: How the Social Observability of an Initial Act of Token Support

Affects Subsequent Prosocial Action." *Journal of Consumer Research* 40, no. 6 (April 1, 2014): 1149–66.

Kroenke, Kurt, Robert L. Spitzer, Janet B. W. Williams, and Bernd Löwe. "An Ultra-Brief Screening Scale for Anxiety and Depression: The PHQ-4." *Psychosomatics* 50, no. 6 (2009): 613–21.

Krysan, Maria, and Sarah Patton Moberg. "Trends in Racial Attitudes." Racial Attitudes Update. Institute of Government and Public Affairs, University of Illinois. August 25, 2016. https://igpa.uillinois.edu/programs/racial-attitudes.

"Ku Klux Klan." Southern Poverty Law Center. Accessed July 10, 2022. www .splcenter.org/fighting-hate/extremist-files/ideology/ku-klux-klan.

Kurzban, Robert, John Tooby, and Leda Cosmides. "Can Race Be Erased? Coalitional Computation and Social Categorization." *Proceedings of the National Academy of Sciences* 98, no. 26 (2001): 15387–92.

Kweli, Talib. *Vibrate Higher: A Rap Story*. New York: Macmillan Audio, 2021.

LaPiere, Richard T. "Attitudes vs. Actions." *Social Forces* 13, no. 2 (1934): 230–37.

Lane, Daniel S., Dam Hee Kim, Slgi S. Lee, Brian E. Weeks, and Nojin Kwak. "From Online Disagreement to Offline Action: How Diverse Motivations for Using Social Media Can Increase Political Information Sharing and Catalyze Offline Political Participation." *Social Media + Society* 3, no. 3 (2017): 2056305117716274.

Lapidot-Lefler, Noam, and Azy Barak. "Effects of Anonymity, Invisibility, and Lack of Eye-Contact on Toxic Online Disinhibition." *Computers in Human Behavior* 28, no. 2 (March 2012): 434–43.

Lee, Jasmine C., and Haeyoun Park. "In 15 High-Profile Cases Involving Deaths of Blacks, One Officer Faces Prison Time." *New York Times* (US edition). May 17, 2017. www.nytimes.com/interactive/2017/05/17/us/black-deaths-police.html.

Lee, Spike. "Eddie Murphy and Spike Lee in Conversation." SPIN, November 19, 2020. https://www.spin.com/featured/eddie-murphy-and-spike-lee-in-conversation-our-1990-cover-story/. Originally published in the October 1990 issue of SPIN.

Leonardo, Zeus, and Ronald K. Porter. "Pedagogy of Fear: Toward a Fanonian Theory of 'Safety' in Race Dialogue." *Race, Ethnicity and Education* 13, no. 2 (July 2010): 139–57.

Leonardo, Zeus, and Michalinos Zembylas. "Whiteness as Technology of Affect: Implications for Educational Praxis." *Equity and Excellence in Education* 46, no. 1 (January 2013): 150–65.

Levenson, Eric, Sarah Jorgensen, Polo Sandoval, and Beech, Samantha. "Mass Shooting at Buffalo Supermarket Was a Racist Hate Crime, Police Say." CNN. May 16, 2022. www.cnn.com/2022/05/15/us/buffalo-supermarket-shooting-sunday/index.html.

Levine, Daniel. *Bayard Rustin and the Civil Rights Movement*. New Brunswick, NJ: Rutgers University Press, 2000.

Levine, Lawrence W. *Black Culture and Black Consciousness: Afro-American Folk Thought from Slavery to Freedom*. Oxford: Oxford University Press, 1978.

Lipsky, Michael. *Street-Level Bureaucracy: The Dilemmas of the Individual in Public Service*. New York: Russell Sage Foundation, 1980.

Lopez, German. "'Ta-Nehisi Coates Has an Incredibly Clear Explanation for Why White People Shouldn't Use the N-Word." Vox. November 9, 2017. www.vox.com/identities/2017/11/9/16627900/ta-nehisi-coates-n-word.

Lorde, Audre. "The Uses of Anger." *Women's Studies Quarterly* 25, no. 1–2 (1997): 278–85.

Loury, G. C. "Self-Censorship in Public Discourse: A Theory of 'Political Correctness' and Related Phenomena." *Rationality and Society* 6, no. 4 (1994): 428–61.

Lundman, Richard J., and Robert L. Kaufman. "Driving While Black: Effects of Race, Ethnicity, and Gender on Citizen Self-Reports of Traffic Stops and Police Actions." *Criminology* 41, no. 1 (2003): 195–220.

Marable, Manning, and Vanessa Agard-Jones, eds. *Transnational Blackness: Navigating the Global Color Line*. Houndmills, UK: Palgrave Macmillan, 2008.

Márquez, John D. "The Black Mohicans: Representations of Everyday Violence in Postracial Urban America." *American Quarterly* 64, no. 3 (2012): 625–51.

Martin, Brittany. "A Look at the 'People's Budget' L.A. Activists Are Promoting." *Los Angeles Magazine* (blog). June 2, 2020. www.lamag.com/citythinkblog/peoples-budget-la-city-budget-lapd/.

Martinez, Theresa A. "The Double-Consciousness of Du Bois and I 'Mestiza Consciousness' of Anzaldúa." *Race, Gender and Class* 9, no. 4 (2002): 158–76.

Massey, Douglas S., and Karen A. Pren. "Origins of the New Latino Underclass." *Race and Social Problems* 4, no. 1 (April 2012): 5–17.

May, Reuben A. Buford. "Velvet Rope Racism, Racial Paranoia, and Cultural Scripts: Alleged Dress Code Discrimination in Urban Nightlife, 2000–2014." *City and Community* 17, no. 1 (2018): 44–64.

McConahay, John B. "Modern Racism, Ambivalence, and the Modern Racism Scale." In *Prejudice, Discrimination, and Racism,* John F. Dovidio and Samuel L. Gaertner, 91–125. Orlando: Academic Press, 1986. http://doi.apa.org/psycinfo /1986-98698-004.

McDowell, Meghan G., and Luis A. Fernandez. "'Disband, Disempower, and Disarm': Amplifying the Theory and Practice of Police Abolition." *Critical Criminology* 26, no. 3 (2018): 373–91.

McIntosh, Peggy. "White Privilege: Unpacking the Invisible Knapsack." *Peace and Freedom Magazine* (July–August 1989): 10–12. https://psychology.umbc.edu /wp-content/uploads/sites/57/2016/10/White-Privilege_McIntosh-1989.pdf.

Meckler, Laura, and Kate Rabinowitz. "The Lines That Divide: School District Boundaries Often Stymie Integration." *Washington Post.* December 16, 2019. www.washingtonpost.com/education/2019/12/16/lines-that-divide-school-district-boundaries-often-stymie-integration/.

Meraz, Sharon. "Is There an Elite Hold? Traditional Media to Social Media Agenda Setting Influence in Blog Networks." *Journal of Computer-Mediated Communication* 14, no. 3 (April 2009): 682–707. https://doi.org/10.1111/j.1083-6101.2009.01458.x.

Miller, Joshua Rhett. "Sorority Sister Speaks Out on Racist Videos That Got Her Booted from College." *New York Post* (blog). January 17, 2018. https://nypost.com/2018/01/17/sorority-sister-booted-for-posting-profanity-laced-racist-videos/.

Miller, Reuben Jonathan. *Halfway Home: Race, Punishment, and the Afterlife of Mass Incarceration.* New York: Little Brown, 2021.

Miller, Ryan W. "NYC Is Sending Social Workers Instead of Police to Some 911 Calls. Here's How It's Going." USA Today. July 23, 2021.www.usatoday.com /story/news/nation/2021/07/23/nyc-mental-health-911-pilot-program-harlem /8053555002/.

Moffatt, Michael. *Coming of Age in New Jersey: College and American Culture.* New Brunswick, NJ: Rutgers University Press, 1989.

Montanaro, Domenico. "President-Elect Joe Biden Hits 80 Million Votes In Year Of Record Turnout." NPR. November 25, 2020. www.npr.org/2020/11/25/937248659 /president-elect-biden-hits-80-million-votes-in-year-of-record-turnout.

Moorwood, Victoria. "Twitter Reacts to Stacey Dash's Apology." *REVOLT* (blog). March 11, 2021. www.revolt.tv/article/2021-03-11/58085/twitter-reacts-to-stacey-dashs-apology/.

Morgan, Kenneth. "George Washington and the Problem of Slavery." *Journal of American Studies* 34, no. 2 (2000): 279–301.

Mundt, Marcia, Karen Ross, and Charla M. Burnett. "Scaling Social Movements through Social Media: The Case of Black Lives Matter." *Social Media + Society* 4, no. 4 (2018): 2056305118807911.

Nadal, Kevin L., Yinglee Wong, Katie E. Griffin, Kristin Davidoff, and Julie Sriken. "The Adverse Impact of Racial Microaggressions on College Students' Self-Esteem." *Journal of College Student Development* 55, no. 5 (July 2014): 461–74.

Nagel, Molly. "Biden Cites Charlottesville and Saving 'Soul' of US in 2020 Presidential Bid." ABC News. April 25, 2019. https://abcnews.go.com/Politics/joe-biden-announces-2020-run-president/story?id=62579665.

Nakamura, Lisa. *Digitizing Race: Visual Cultures of the Internet.* New ed. Minneapolis: University of Minnesota Press, 2008.

Nellis, Ashley. "The Color of Justice: Racial and Ethnic Disparity in State Prisons." The Sentencing Project. October 13, 2021. https://www.sentencing-project.org/publications/color-of-justice-racial-and-ethnic-disparity-in-state-prisons/.

Ng, Jennifer C., Sharon S. Lee, and Yoon K. Pak. "Contesting the Model Minority and Perpetual Foreigner Stereotypes: A Critical Review of Literature on Asian Americans in Education." *Review of Research in Education* 31, no. 1 (2007): 95–130.

Noble, Safiya Umoja. *Algorithms of Oppression: How Search Engines Reinforce Racism.* New York: New York University Press, 2018.

Noble, Safiya Umoja, and Brendesha M. Tynes. *The Intersectional Internet: Race, Sex, Class, and Culture Online.* New York: Peter Lang, 2016.

Nobles, Ryan. "Marjorie Taylor Greene Compares House Mask Mandates to the Holocaust." CNN Politics. May 22, 2021. www.cnn.com/2021/05/21/politics/marjorie-taylor-greene-mask-mandates-holocaust/index.html.

Nunnally, Shayla C., and Niambi M. Carter. "Moving from Victims to Victors: African American Attitudes on the 'Culture of Poverty' and Black Blame." *Journal of African American Studies* 16, no. 3 (2012): 423–55.

Pager, Devah, Bart Bonikowski, and Bruce Western. "Discrimination in a Low-Wage Labor Market: A Field Experiment." *American Sociological Review* 74, no. 5 (October 1, 2009): 777–99.

Palos, Ari Luis, dir. *Precious Knowledge*. Arizona: Dos Vatos Productions, 2011. www.pbs.org/independentlens/films/precious-knowledge/.

Parent, Wayne. "A Liberal Legacy: Blacks Blaming Themselves for Economic Failures." *Journal of Black Studies* 16, no. 1 (1985): 3–20.

Payne, Charles M. *I've Got the Light of Freedom: The Organizing Tradition and the Mississippi Freedom Struggle*. Oakland: University of California Press, 2007.

Peebles, Maurice. "7 Common Sense Reasons Why College Athletes Should Be Paid (According to Jay Bilas)." Complex. December 3, 2015. www.complex.com/sports/2015/12/jay-bilas-interview/.

Perez, Chris. "White Man Calls Police on Black Family at Neighborhood Pool." *New York Post*. July 5, 2018. https://nypost.com/2018/07/05/white-man-calls-police-on-black-family-at-neighborhood-pool/.

Pérez Huber, Lindsay, and Daniel G. Solorzano. "Racial Microaggressions as a Tool for Critical Race Research." *Race, Ethnicity and Education* 18, no. 3 (May 4, 2015): 297–320.

"Personal Privilege Profile." Interaction Institute for Social Change. 2011. https://foodsolutionsne.org/wp-content/uploads/2015/02/PERSONAL_PRIVILEGE_PROFILE.pdf.

Petersilia, Joan. "Racial Disparities in the Criminal Justice System: A Summary." *Crime and Delinquency* 31, no. 1 (January 1, 1985): 15–34.

Picca, Leslie Houts, and Joe R. Feagin. *Two-Faced Racism: Whites in the Backstage and Frontstage*. New York: Routledge, 2007.

Pierce, Chester. "Offensive Mechanisms." In *The Black Seventies*, edited by Flyod B. Barbour, 265–82. Boston: Porter Sargent, 1970.

Pieterse, Alex L., and Robert T. Carter. "The Role of Racial Identity in Perceived Racism and Psychological Stress Among Black American Adults: Exploring Traditional and Alternative Approaches." *Journal of Applied Social Psychology* 40, no. 5 (May 2010): 1028–53.

Pieterse, Alex L., Nathan R. Todd, Helen A. Neville, and Robert T. Carter. "Perceived Racism and Mental Health among Black American Adults: A Meta-Analytic Review." *Journal of Counseling Psychology* 59, no. 1 (2012): 1–9.

Pollock, Mica. *Colormute: Race Talk Dilemmas in an American School*. Princeton, NJ: Princeton University Press, 2004.

Potok, Mark. "The 'Patriot' Movement Explodes." Southern Poverty Law Center. Intelligence Report. March 1, 2012. www.splcenter.org/fighting-hate /intelligence-report/2012/patriot-movement-explodes.

Prather, Cynthia, Taleria R. Fuller, Khiya J. Marshall, and William L. Jeffries IV. "The Impact of Racism on the Sexual and Reproductive Health of African American Women." *Journal of Women's Health* 25, no. 7 (2016): 664–71.

"Presidential Approval Ratings—Donald Trump." Gallup. November 16, 2016. https://news.gallup.com/poll/203198/presidential-approval-ratings-donald-trump.aspx.

Pyke, Karen D. "What Is Internalized Racial Oppression and Why Don't We Study It? Acknowledging Racism's Hidden Injuries." *Sociological Perspectives* 53, no. 4 (2010): 551–72.

Ransby, Barbara. *Ella Baker and the Black Freedom Movement: A Radical Democratic Vision*. Chapel Hill: University of North Carolina Press, 2003.

Rascoe, Ayesha. "Why Are White Nationalist Groups Targeting LGBTQ Groups?" NPR. June 19, 2022. www.npr.org/2022/06/19/1106125400/why-are-white-nationalist-groups-targeting-lgbtq-groups.

Rashid, Neha. "The Emergence of the White Troll behind a Black Face." NPR Code Switch. March 21, 2017. www.npr.org/sections/codeswitch/2017/03/21 /520522240/the-emergence-of-the-white-troll-behind-a-black-face.

Ray, Rashawn, Melissa Brown, Neil Fraistat, and Edward Summers. "Ferguson and the Death of Michael Brown on Twitter: #BlackLivesMatter, #TCOT, and the Evolution of Collective Identities." *Ethnic and Racial Studies* 40, no. 11 (2017): 1797–813.

Reilly, Katie. "Donald Trump Blames 'Both Sides' for Charlottesville Clashes." *Time*. August 15, 2017. https://time.com/4902129/president-donald-trump-both-sides-charlottesville/.

Rhodes, Jesse H. "Voting Rights Politics in the Age of Obama, 2009–2016." In *Ballot Blocked*, 163–80. Stanford, CA: Stanford University Press, 2020.

Richardson, Allissa V. *Bearing Witness While Black: African Americans, Smartphones, and the New Protest #Journalism*. New York: Oxford University Press, 2020.

———. "Bearing Witness While Black: Theorizing African American Mobile Journalism after Ferguson." *Digital Journalism* 5, no. 6 (2017): 673–98.

Richardson, Lynne D., and Marlaina Norris. "Access to Health and Health Care: How Race and Ethnicity Matter." *Mount Sinai Journal of Medicine: A Journal of Translational and Personalized Medicine* 77, no. 2 (2010): 166–77.

Ritchie, Andrea, Deirdre Smith, Janetta Johnson, Jumoke Ifetayo, Marbre Stahly-Butts, Montague Simmons, Nkechi Taifa, Rachel Herzing, Richard Wallace, and Taliba Obuya. "Reparations Now Toolkit." M4BL. 2020. https://m4bl.org/wp-content/uploads/2020/05/Reparations-Now-Toolkit-FINAL.pdf.

Rogers, Richard A. "From Cultural Exchange to Transculturation: A Review and Reconceptualization of Cultural Appropriation." *Communication Theory* 16, no. 4 (2006): 474–503.

Rosario, Isabella. "The Unlikely Story Behind Japanese Americans' Campaign for Reparations." NPR Code Switch. March 24, 2020.www.npr.org/sections /codeswitch/2020/03/24/820181127/the-unlikely-story-behind-japanese-americans-campaign-for-reparations.

Rose, Joel. "The Inquiry into Border Agents on Horseback Continues. Critics See a 'Broken' System." NPR. November 6, 2021. www.npr.org/2021/11/06 /1052786254/border-patrol-agents-horseback-investigation-haitian-immigrants.

Ross, Cody T., Bruce Winterhalder, and Richard McElreath. "Racial Disparities in Police Use of Deadly Force against Unarmed Individuals Persist after Appropriately Benchmarking Shooting Data on Violent Crime Rates." *Social Psychological and Personality Science* 12, no. 3 (2021): 323–32.

Rothstein, Richard. *The Color of Law: A Forgotten History of How Our Government Segregated America.* New York: Liveright, 2017.

Rowan-Kenyon, Heather T., Ana M. Martínez Alemán, Kevin Gin, Bryan Blakeley, Adam Gismondi, Jonathan Lewis, Adam McCready, Daniel Zepp, and Sarah Knight. "Social Media in Higher Education." *ASHE Higher Education Report* 42, no. 5 (2016): 7–128.

Rundle, Margaret, Emily Weinstein, Howard Gardner, and Carrie James. "Doing Civics in the Digital Age: Casual, Purposeful, and Strategic Approaches to Participatory Politics." Youth and Participatory Politics Research Network. September 30, 2017. https://ypp.dmlcentral.net/publications/238.

Ryan, Josiah. "'This Was a Whitelash': Van Jones' Take on the Election Results." CNN Politics. November 9, 2016. www.cnn.com/2016/11/09/politics/van-jones-results-disappointment-cnntv/.

Schaffner, Brian F., Matthew MacWilliams, and Tatishe Nteta. "Explaining White Polarization in the 2016 Vote for President: The Sobering Role of Racism and Sexism." Paper presented at the conference on The U.S. Elections of 2016: Domestic and International Aspects, 2017. IDC Herzliya. http://people .umass.edu/schaffne/schaffner_et_al_IDC_conference.pdf

Schulman, Kevin A., Jesse A. Berlin, William Harless, Jon F. Kerner, Shyrl Sistrunk, Bernard J. Gersh, Ross Dubé, et al. "The Effect of Race and Sex on Physicians' Recommendations for Cardiac Catheterization." *New England Journal of Medicine* 340, no. 8 (February 25, 1999): 618–26. https://doi.org/10.1056 /NEJM199902253400806.

Schwartz, Sarah. "Map: Where Critical Race Theory Is Under Attack." Education Week. June 11, 2021. www.edweek.org/policy-politics/map-where-critical-race-theory-is-under-attack/2021/06.

Sela, Shlomi, Tsvi Kuflik, and Gustavo S. Mesch. "Changes in the Discourse of Online Hate Blogs: The Effect of Barack Obama's Election in 2008." *First Monday* 17, no. 11 (2012). https://journals.uic.edu/ojs/index.php/fm/article /view/4154.

Sellers, Robert M., and J. Nicole Shelton. "The Role of Racial Identity in Perceived Racial Discrimination." *Journal of Personality and Social Psychology* 84, no. 5 (2003): 1079–92.

Sellers, Robert M., Mia A. Smith, J. Nicole Shelton, Stephanie A. J. Rowley, and Tabbye M. Chavous. "Multidimensional Model of Racial Identity: A Reconceptualization of African American Racial Identity." *Personality and Social Psychology Review* 2, no. 1 (1998): 18–39.

Sellers, Robert M., Nikeea Copeland-Linder, Pamela P. Martin, and R. L'Heureux Lewis. "Racial Identity Matters: The Relationship between Racial Discrimination and Psychological Functioning in African American Adolescents." *Journal of Research on Adolescence* 16, no. 2 (2006): 187–216.

"Settling For Misconduct: Police Lawsuits in Chicago." Chicago Reporter. Accessed June 26, 2022. http://projects.chicagoreporter.com/settlements.

Siegel, Rachel. "Two Black Men Arrested at Starbucks Settle with Philadelphia for $1 Each." *Washington Post.* May 3, 2018. www.washingtonpost.com/news /business/wp/2018/05/02/african-american-men-arrested-at-starbucks-reach-1-settlement-with-the-city-secure-promise-for-200000-grant-program-for-young-entrepreneurs/.

Simes, Jessica T. *Punishing Places: The Geography of Mass Imprisonment*. Oakland: University of California Press, 2021.

Singal, Jesse. "Psychology's Favorite Tool for Measuring Racism Isn't Up to the Job." The Cut. January 11, 2017. www.thecut.com/2017/01/psychologys-racism-measuring-tool-isnt-up-to-the-job.html.

Small, Mario Luis. "'How Many Cases Do I Need?' On Science and the Logic of Case Selection in Field-Based Research." *Ethnography* 10, no. 1 (2009): 5–38.

Smith, Clint. *How the Word Is Passed: A Reckoning with the History of Slavery Across America*. New York: Little, Brown, 2021.

Smith, William A. *Racial Battle Fatigue in Higher Education: Exposing the Myth of Post-Racial America*. Lanham, MD: Rowman & Littlefield, 2014.

Smock, Kristina. *Democracy in Action: Community Organizing and Urban Change*. New York: Columbia University Press, 2004.

Snow, Aurora. "Black Porn Stars Come Forward with Their Racism Horror Stories." The Daily Beast. June 6, 2020. www.thedailybeast.com/black-porn-stars-come-forward-with-their-racism-horror-stories.

———. "The Rise of Racist Porn." The Daily Beast. June 23, 2018. www.thedailybeast.com/the-rise-of-racist-porn.

Solorzano, Daniel, Miguel Ceja, and Tara Yosso. "Critical Race Theory, Racial Microaggressions, and Campus Racial Climate: The Experiences of African American College Students." *Journal of Negro Education* 69, no. 1–2 (2000): 60–73.

Spencer, Steven J., Claude M. Steele, and Diane M. Quinn. "Stereotype Threat and Women's Math Performance." *Journal of Experimental Social Psychology* 35, no. 1 (January 1999): 4–28.

Steele, Catherine Knight. *Digital Black Feminism*. New York: New York University Press, 2021.

Steele, Claude M., and Joshua Aronson. "Stereotype Threat and the Intellectual Test Performance of African Americans." *Journal of Personality and Social Psychology* 69, no. 5 (1995): 797–811.

Steen, Sara, Rodney L. Engen, and Randy R. Gainey. "Images of Danger and Culpability: Racial Stereotyping, Case Processing, and Criminal Sentencing." *Criminology* 43, no. 2 (2005): 435–68.

Steffensmeier, Darrell, and Stephen Demuth. "Ethnicity and Sentencing Outcomes in US Federal Courts: Who Is Punished More Harshly?" *American Sociological Review* 65, no. 5 (2000): 705–29.

Stephens-Davidowitz, Seth. "The Cost of Racial Animus on a Black Candidate: Evidence Using Google Search Data." *Journal of Public Economics* 118 (October 2014): 26–40.

Stolworthy, Jacob. "Kendrick Lamar Explains Why He Criticised White Fan Who Rapped the N-Word." Independent. June 29, 2018. www.independent .co.uk/arts-entertainment/music/news/kendrick-lamar-white-fan-rap-n-word-racism-concert-explained-a8423101.html.

Stone, Geoffrey R. "Aims of Education Address 2016." YouTube video, 55:56. University of Chicago, September 28, 2016. www.youtube.com/watch?v= lfhRm6hNIYw.

Sue, Derald Wing. *Microaggressions in Everyday Life: Race, Gender, and Sexual Orientation.* Hoboken, NJ: Wiley, 2010.

Thomas, James M. "Du Bois, Double Consciousness, and the 'Jewish Question.'" *Ethnic and Racial Studies* 43, no. 8 (2020): 1333–56.

Thompson, Julian G., and Rob Eschmann. "Tweeting Abolition in an Age of Mass Incarceration and Social Unrest, Part I: What Is Abolition?" *Sociology Lens*(blog), October 10, 2022. https://www.sociologylens.net/topics/collective-behavior-and-social-movements/tweeting-abolition-mass-incarceration-part-i /42880.

Timberg, Craig, and Elizabeth Dwoskin. "Proud Boys, Right-Wing Extremists Celebrate Trump's Failure to Condemn White Supremacy." *Washington Post.* September 30, 2020. www.washingtonpost.com/technology/2020/09 /30/trump-debate-rightwing-celebration/.

Timm, Jane C. "Trump Says Obama Didn't Reform Policing—but He Did. Then the President Ditched It." June 16, 2020. www.nbcnews.com/politics/donald -trump/trump-says-obama-didn-t-reform-policing-he-did-then-n1231200.

Timmermans, Stefan, and Iddo Tavory. "Theory Construction in Qualitative Research from Grounded Theory to Abductive Analysis." *Sociological Theory* 30, no. 3 (2012): 167–86.

Tobar, Hector. "Deputies in 'Neo-Nazi' Gang, Judge Found: Sheriff's Department: Many at Lynwood Office Have Engaged in Racially Motivated Violence against Blacks and Latinos, Jurist Wrote." *Los Angeles Times.* October 12, 1991. www.latimes.com/archives/la-xpm-1991-10-12-me-107-story.html.

Tolan, Casey. "DC Police Made Far More Arrests at Height of Black Lives Matter Protests than during Capitol Clash." CNN. January 9, 2021.

www.cnn.com/2021/01/08/us/dc-police-arrests-blm-capitol-insurrection-invs /index.html.

Trent, Maria, Danielle G. Dooley, and Jacqueline Dougé. "The Impact of Racism on Child and Adolescent Health." *Pediatrics* 144, no. 2 (2019): e20191765.

Tripodi, Francesca. "Yakking about College Life: Examining the Role of Anonymous Forums on Community Identity Formation." In *Digital Sociologies,* edited by Jessie Daniels, Karen Gregory, and Tressie McMillan Cottom, 251–70. Bristol, UK: Policy Press, 2016.

Ture, Kwame, and Charles V. Hamilton. *Black Power: The Politics of Liberation in America.* New York: Vintage, 1992.

Twain, Mark. *The Adventures of Tom Sawyer and the Adventures of Huckleberry Finn.* Modern Library of the World's Best Books. New York: Modern Library, 1922.

Tynes, Brendesha M., Adriana J. Umana-Taylor, Chad A. Rose, Johnny Lin, and Carolyn J. Anderson. "Online Racial Discrimination and the Protective Function of Ethnic Identity and Self-Esteem for African American Adolescents." *Developmental Psychology* 48, no. 2 (2012): 343–55.

Uebele, Hannah. "All Rev'd Up: Saying That America Isn't Racist Is 'Absurd.'" GBH News. May 3, 2021. www.wgbh.org/news/national-news/2021/05/03 /all-revd-up-saying-that-america-isnt-racist-is-absurd.

"Unsolved Mysteries: Do Any Republicans Know What Critical Race Theory Actually Is?" The Daily Show with Trevor Noah. June 29, 2021. YouTube video, 3:40. www.youtube.com/watch?v=6ofjZH8oy3g.

Vagianos, Alanna. "Jemele Hill n Calling Trump a White Supremacist: 'I Thought I Was Saying Water Is Wet.'" HuffPost. December 27, 2018. www .huffpost.com/entry/jemele-hill-trump-white-supremacist-water-is-wet_n_ 5c24f8b0e4b05c88b6fe3477.

Velshi, Ali. "Robin Rue Simmons: 'It Must Start. That's How You Seek Reparations. You Start.'" MSNBC.com, June 19, 2022. www.msnbc.com/ali-velshi /watch/robin-rue-simmons-it-must-start-that-s-how-you-seek-reparations-you-start-142419013980.

Viala-Gaudefroy, Jérôme, and Dana Lindaman. "Donald Trump's 'Chinese Virus': The Politics of Naming." The Conversation. April 21, 2020. http:// theconversation.com/donald-trumps-chinese-virus-the-politics-of-naming-136796.

Wang, Qun. "'Double Consciousness,' Sociological Imagination, and the Asian American Experience." *Race, Gender and Class* 4, no. 3 (1997): 88–94.

Washington, Jesse. "We Finally Have Answers about Michael Jordan and 'Republicans Buy Sneakers, Too.'" Andscape. May 4, 2020. https://andscape .com/features/we-finally-have-answers-about-michael-jordan-and-republicans-buy-sneakers-too/.

West, Rebecca J., and Bhoomi K. Thakore. "Racial Exclusion in the Online World." *Future Internet* 5, no. 2 (2013): 251–67.

"What Does Free Speech Mean?" United States Courts. Accessed February 12, 2021. www.uscourts.gov/about-federal-courts/educational-resources/about -educational-outreach/activity-resources/what-does.

Williams, Apryl. "Black Memes Matter: #LivingWhileBlack with Becky and Karen." *Social Media + Society* 6, no. 4 (October 1, 2020): 2056305120981047.

Williams, David R. "How Racism Makes Us Sick." TED Talk video, 17:19. 2016. www.ted.com/talks/david_r_williams_how_racism_makes_us_sick /transcript.

Williams, David R., and Chiquita Collins. "Racial Residential Segregation: A Fundamental Cause of Racial Disparities in Health." *Public Health Reports* 116 (September–October 2016): 404–16. https://www.ncbi.nlm.nih.gov/pmc /articles/PMC1497358/.

Williams, David R., Yan Yu, James S. Jackson, and Norman B. Anderson. "Racial Differences in Physical and Mental Health: Socio-Economic Status, Stress and Discrimination." *Journal of Health Psychology* 2, no. 3 (1997): 335–51.

Wilson, William Julius. *Power, Racism and Privilege: Race Relations in Theoretical and Sociohistorical Perspectives.* New York: Macmillian., 1973.

Wise, Alana. "Juneteenth Is Now a Federal Holiday." NPR. June 17, 2021.www .npr.org/2021/06/17/1007602290/biden-and-harris-will-speak-at-the-bill-signing-making-juneteenth-a-federal-holi.

Wu, Nicholas. "Cory Booker Denounces White Supremacy at Charleston Church Shooting Site." USA Today. August 7, 2019. www.usatoday.com /story/news/politics/2019/08/07/cory-booker-denounces-white-supremacy-charleston-church-shooting-site/1942497001/.

Yam, Kimmy. "Viral Images Show People of Color as Anti-Asian Perpetrators. That Misses the Big Picture." NBC News. June 15, 2021. www.nbcnews .com/news/asian-america/

viral-images-show-people-color-anti-asian-perpetrators-misses-big-n1270821.

———. "Anti-Asian Hate Crimes Increased 339 Percent Nationwide Last Year, Report Says." NBC News. January 31, 2022. www.nbcnews.com/news /asian-america/anti-asian-hate-crimes-increased-339-percent-nationwide-last-year-repo-rcna14282.

Yosso, Tara, William Smith, Miguel Ceja, and Daniel Solórzano. "Critical Race Theory, Racial Microaggressions, and Campus Racial Climate for Latina/o Undergraduates." *Harvard Educational Review* 79, no. 4 (2009): 659–91.

Yousef, Odette. "31 Members of a White Nationalist Group Were Arrested for Planning to Riot at Pride." NPR. June 12, 2022. www.npr.org/2022/06/12 /1104418170/31-members-of-a-white-nationalist-group-were-arrested-for-planning-to-riot-at-pr.

Zhang, Yan, Lening Zhang, and Francis Benton. "Hate Crimes against Asian Americans." *American Journal of Criminal Justice* 47 (2022): 441–61.

Index

Crenshaw, Kimberlé, 195

crimes, anti-Asian hate, 16–17, 43, 175, 206–7

criminality, assumptions of, 109–10, 123–25

criminal justice system: prison abolitionists on, 160, 198*fig.*, 199–200; racial disparities in, 40, 58–59, 102–3, 122, 199

critical race theory (CRT): bans on, 53–54, 58, 162; as catchphrase for antiracism, 251n5; defined, 53; misunderstandings of, 53–54

cultural appropriation, 163, 194, 196*fig.*

Cuomo, Chris, 23–24

Cyrus, Miley, 163

Daily Show with Trevor Noah, The, 53–54

dancing: Black, cultural appropriation of, 163; White, stereotypes of, 152–53

Daniels, Jessie, 6, 17, 45

dapping, 106, 245n3

Dash, Stacy, 161–62

dating, interracial, 25

Davis, Angela, 199, 209

de facto segregation, 37

defensiveness, of White people in racial discourse, 75–76, 127, 133

democracy: threats in 2021 to, 26–27; White supremacy and, 212

Democratic Party, Google searches for n-word and, 44–45

deviance, politics of, 144, 154–55

DiAngelo, Robin, 75

digital dualism fallacy, 8

digital rage, 45

Dillard, Sabriya, 228

discomfort of White people, in rules of racial discourse, 74–79

disinhibition, 6

disrespectability politics, 145

disruption, as goal of nonviolence, 48–49

dog whistle, 197, 198*fig.*

double consciousness, 136–58; applied to colonized peoples, 140; contemporary scholarship on, 140–41; defined, 18–19, 136, 139–40; vs. double-sided consciousness, 141–42; Du Bois's individual-level solution to, 140, 150, 156; microaggressions and, 155; moving beyond, 18–19, 149–52; public regard in, 154; respectability politics in, 144–45; second-sight in, 136, 145, 148, 155; stereotype threat and, 152; veiled speech and, 150–52; veil in, 139–40, 145, 149–50; Wypipo tweets as response to, 142–49

double-sided consciousness, 136–58; awareness of racism in, 152–56; defined, 141–42; externalizing of racism in, 147–48, 154, 157; as public and communal project, 141–42, 147; second-sight in, 148–49; veiled speech in, 151–52. *See also* Wypipo tweets

Douglass, Frederick, 204

Du Bois, W. E. B.: on color line problem, 140–41; on second-sight, 136, 145, 148, 155; *The Souls of Black Folk,* 136, 150; on veil, 139–40, 145, 149–50. *See also* double consciousness

Duke, David, 24, 188

Durant, Kevin, 247n3

echo chambers, 183

education: academic costs of activism in, 171–72, 174–75; achievement gap in, 40,

invasive species, 57
iPhones, 29
Israel: in conflict with Palestinians, 208–11; German reparations paid to, 201

Jackson, Andrew, 130
Jackson, Sarah, 13, 51, 112, 175–76
James, Lebron, 93
January 6, 2021, Capitol riots, 17, 26–27, 63, 212
Japanese American reparations, 201–2
Jefferson, Thomas, 55–56
Jemisin, N. K., 20, 45
Jim Crow era: contemporary racist ideology vs., 6; decrease in visibility of racism after, 15–16; defined, 239n3; Ku Klux Klan in, 15; racist attitudes in justification of, 36–37; racist language during, 6; reparations for, 203
jokes, racist, 34, 46, 88
Jones, Anthony Van, 187
Jones, Feminista, 194
Jones, Leslie Kay, 156
Jordan, Michael, 93–94, 244nn20,22
Jordan, Robert, 104, 131
Jordan brand shoes, 22, 244nn20,22
journalism. See media coverage
judges, 58–59, 102–3
Juneteenth, 203–4, 205fig.
Jurgenson, Nathan, 8

Kaepernick, Colin, 93, 134, 214
Kahne, James, 176
Kendi, Ibram X., 55, 65–66
Kerry, John, 44
Key and Peele, 245n3

King, Don, 213
King, Martin Luther, Jr., 45, 55, 146, 159, 167, 168, 195
Kluegel, James, 63
Kronbach, Kris, 23
Ku Klux Klan (KKK), 15, 24
Kwanzaa, 204
Kweli, Talib, 209

laboratory studies, on implicit bias, 32–34
laissez-faire racism, 38, 63–64
Lamar, Kendrick, 136–38, 247n1, 248n21
language, antiracist, 65–68, 192–94. See also racist language; Twitter, race-related terms on
Latino threat narrative, 38
Latinx people: common themes of microaggressions toward, 110; immigration by, 38, 201; in racial and ethnic studies classes, 55; racial violence against, 17; social media use by, 12
Lawrence, Jennifer, 95
leadership: in civil rights movement, 167–68, 195–97; organizing as commitment to developing, 168
Lee, Opal, 204
Lee, Robert E., 204
Lee, Spike, 213
Leonardo, Zeus, 64, 173
Levine, Lawrence W., 150
Lipsky, Michael, 69
Lorde, Audre, 72, 77, 78, 186, 195
Lord of Chaos (Jordan), 104

"m.A.A.d city" (song), 136–37, 247n1
Macklemore, 151, 248n21

New Jim code, 46
New Jim Crow, The (Alexander), 59, 199
New Racism, 245n4
news media. See media coverage
Ng, Nigel, 153
Nike, 134, 214
Noble, Safiya Umoja, 46, 51
Non-Black People of Color (NBPOC), 145
nonverbal communication, online, 44
nonviolence, in civil rights movement, 48–50, 159
n-word: Black use of, 68; college campus policies on, 68–72; Google searches for, 44–45; as microaggression, 72; in music, 4–5, 72, 136–39; online vs. in real world, 3–5; in video game chats, 1–3

Obama, Barack: on Clinton as first Black president, 153; dapping by, 245n3; Google searches for n-word and, 44–45; immigration policy of, 201; in myth of postracial era, 186–87; policing taskforce under, 56; on presidential election of 2016, 187; racist backlash to presidency of, 187
Oculus, 7
Odinet, Michelle, 102–3
offending White people, in rules of racial discourse, 75–79
old-fashioned racism. See overt racism
open racism. See overt racism
organizing, 167–75; in civil rights movement, 168–69, 197; on college campuses, limitations of, 170–75; community, 185; defined, 168; models of, 169–70; in Movement for Black Lives, 169; online, outcomes of, 177
othering, of Whiteness, 142–43

overt (old-fashioned) racism: decline of, 106; in definitions of racism, 27, 162; at January 6 Capitol riots, 17; media coverage of, 27; racist label restricted to, 26, 86, 109, 191; surveys measuring, 30; of Trump, 188. See also racism, visibility of

Palestine: impact of videos from, 164–65, 208–9, 216; use of term on Twitter, 208–11, 210fig.
pandemic. See COVID-19 pandemic
Parks, Rosa, 167, 168–69, 195–97
participatory politics, 176–85
patriarchy, 195, 196fig.
Patterson, Robert, 202
Payne, Charles, 49, 74–75, 167, 168
PC. See political correctness
Peele, Jordan, 151
PIC. See Politically Incorrect Confessions
Picca, Leslie Houts, 34, 86
Pierce, Chester, 107
"Pilgrimage to Nonviolence" (King), 159
plagiarism, 156
police: abolishing or defunding, 200–201, 205fig.; at Black Lives Matter vs. White supremacist rallies, 62–63; called on Black people by White people, 51, 101–2, 125; colorblind racism of, 43; dangers of offending, 76–77; implicit bias of, 32–33; Obama's taskforce on, 56–57
police violence, anti-Black: accountability for, lack of, 179–80, 217; college campuses affected by, 9; gaps in data on, 56–57; Jordan (Michael) on, 93–94; media coverage of, 48; police talking online about, 82–83; protests of 2020

against, 51, 166, 191, 200; and use of term "racism" on Twitter, 191; video recordings of, 48, 99; visibility of racism in, 83

political correctness (PC): in informal rules of racial discourse, 63–68, 81; opposition to, 63–64, 81–82; origins and meaning of term, 63; Politically Incorrect Confessions on, 86

Politically Incorrect Confessions (PIC), 83–103; anonymity of, 84–91, 96–97; content of, 85–91; effects of and responses to, 91–100; number of posts to, 84; purpose of, 84, 85

politics: coded racist language in, 81–82; of deviance, 144, 154–55; dog-whistle, 197; participatory, 176–85; respectability, 144–45. *See also* conservative politics; *specific parties and politicians*

Pollock, Mica, 81

pornography, 153

Porter, Ronald, 173

postracial era, myth of, 186–87, 213

power-based model of organizing, 169

power dynamics: in cultural appropriation, 194; racial, effects of online resistance on, 129–33

prejudice: defined, 32; implicit, 32–34. *See also* bias; racism

Pren, Karen A., 38

presidential elections: of 2004, 44–45; of 2008, 44–45; of 2016, 23–24, 187–88, 191; of 2020, 25–26, 188, 217

press coverage. *See* media coverage

Princeton University, 61

prison: abolitionists on, 160, 198*fig.,* 199–200; racial disparities in, 40, 58–59, 102–3, 199

prison-industrial complex, 160, 199

Pritzker, J. B., 199

privacy, online, 46–47

private vs. public spaces: internet as, 46–47; racist jokes in, 34, 46, 88; two-faced racism theory on, 34–35, 86–87

privilege: as antiracist buzzword, 66; self-recognition of, 92–94, 194. *See also* White privilege

Proud Boys, 62–63

pseudonyms, use of, 12, 228, 248n21

public regard, in racial identity, 154

QR codes, 165, 249n7

race, talking about. *See* racial discourse

racial and ethnic identity: in coping with racism, 141; dimensions of, 153–54; informal online sources of learning about, 162–63

racial and ethnic studies, bans on, 55–56

racial battle fatigue, 155

racial checking, online: defined, 124; effectiveness of, 124–25, 128–32, 134

racial discourse, in-person, informal rules of, 53–80; 1) don't talk about race, 55–62; 2) if you must talk about race, stick to the niceties, 62–68; 3) White people are allowed to break the rules, 68–74; 4) don't make White people uncomfortable, 74–79; effects of Politically Incorrect Confessions on, 99–100; vs. online norms, 79–80; Wypipo tweets breaking, 147–48

racial discourse, online: benefits of, 114–17, 214–15; characteristics of, 41–47, 111; in counterspaces, 112–15; forms of,

racial discourse *(continued)*
41; initiation of, 111–12; norms of,
79–80, 114–15, 118, 125; prevalence of,
111. *See also specific forms and platforms*
racial inequality: Black views on causes
of, 60, 93; Black women in rise of
online attention to, 194–95; colorblind
racism on causes of, 39–40; conservative views on causes of, 162; in
definition of racism, 27, 35, 40;
function of racism as maintenance of,
27, 35–38, 106–7
racial power dynamics, effects of online
resistance on, 129–33
racial slurs. *See* hate speech; racist
language
racial socialization, 7, 154
racial violence, against Asian Americans,
16–17, 43, 175, 206–8. *See also* police
violence
racism: myth of end of, 186–87, 213; use of
term on Twitter, 191–92, 193*fig.*, 230,
234–35*fig. See also specific forms,
locations, and platforms*
racism, admissions of: vs. acts of racism,
20–27; definition of racism based on,
23–26, 109; by police, 83; in surveys,
20, 23, 25, 30–32
racism, definitions of: admission vs. acts
of racism in, 23–26, 109; overt and
covert racism in, 27, 162; racial
inequality as function of racism in, 27,
35, 40; in Wypipo tweets, 148–49
racism, expressions of: importance of
interpersonal challenges to, 215–16;
racial socialization in preparation for,
7, 154; recent normalization of, 16–17.
See also racist language

racism, identifying: approaches to, 29–41;
challenges of, 29, 106–11; criteria for,
24–26. *See also* racist label
racism, resistance to. *See* activism;
antiracism; resistance
racism, talking about. *See* racial
discourse
racism, visibility of: in anti-Black police
violence, 83; decrease after Jim Crow
era, 15–16; need for unmasking to
increase, 103, 189–90; online, ubiquity
of, 82–83; after Trump's election,
187–89
"racism without racists," 6, 86
racist attitudes: effects of nonviolence on,
48–50; historical origins of, 35–36;
racial inequality as function of, 27,
35–38; structural analysis of, 27, 35–38
racist jokes, 34, 46, 88
racist label: admission of racism as
criteria for, 23–27, 109; for calling
police on Black people, 101; as
character assassination, 86; Politically
Incorrect Confessions on, 85–87;
resistance to using, 109; restricted to
overt racism, 26, 86, 109, 191; silence
and PC language as safeguard from,
81; for Trump, 24–25; use on Twitter,
191, 193*fig.*, 234–35*fig.*
racist language: coded, in politics, 81–82;
college campus policies on, 68–74, 101;
in colorblind racism, 6, 100; consequences online, 43–44; function of, as
legitimizing racial inequality, 106–7;
about immigration, 38; Jim Crow vs.
contemporary, 6; microaggressions as
replacement for, 107; and nature of
racism online vs. in real world, 3–5,

213; political correctness and, 64, 66; racial socialization in preparation for, 7; taboos on, 10, 15, 25; of Trump, 9, 16, 95, 175; in video game chats, 1–3, 7. *See also* n-word; *specific platforms*

racist robots, 46

radicalization, 132–33

rage: digital, 45; White, 187

Ramsey, Gordon, 153

Ray, Rashawn, 13, 51, 145, 176

Reconstruction, 15, 237n3

Reddit, 12

Red Scare, 63

reparations: defined, 201; historical examples of, 201–2; and informal rules of racial discourse, 61; by universities, 61; use of term on Twitter, 199, 201–3, 205*fig.*

Republican Party: Google searches for n-word and, 44–45; Trump in future of, 189; voter suppression and, 199. *See also* conservative politics

resistance to racism, in-person: challenges of identifying racism and, 106–11; choosing battles in, 71, 117, 120; dangers of, 148; difficulty of, 125–29; over ID cards on college campuses, 104–9, 119–20, 126; mental health effects of, 131, 229–30, 233–34*fig.*; reasons for importance of, 119–25; reasons for not responding, 107, 115–16, 124, 127

resistance to racism, online, 104–35; on Black Twitter, 51, 144, 157–58; by celebrities, 214; collective, 116–19; in counterspaces, 112–15; dangers of, 132–34; and double-consciousness problem, 141; effectiveness of, 124–25,

128–31, 134, 166; increased agency through, 113, 128–29, 132; mental health effects of, 131, 229–30, 233–34*fig.*; new opportunities for, 47–48, 131–35; vs. organized activism, 157; racial checking in, 124–25, 128–32; racial power dynamics changed by, 129–33; reasons for importance of, 119–25; as unintended consequence of online racism, 18. *See also* activism; antiracism; *specific platforms*

respectability politics, 144–45

restaurants: admitting vs. acting on racism at, 20, 23, 25; civil rights sit-ins at, 48–49

retention rates, 121

reverse racism, 75

Richardson, Allissa, 48

robots: racist, 46; on Twitter (bots), 42, 243n49

Rustin, Bayard, 146, 197

safe spaces, 72–73, 86, 173

sanctioning, informal mechanisms of, 130–32

school segregation, 28, 37–38

search engines: implicit bias in, 46; n-word in, 44–45

search syntax, 227, 236–37

second-sight, 136, 145, 148–49, 155

security, on college campuses, 104–9, 119–20, 126

segregation: racist attitudes in justification of, 36–37; school, 28, 37–38

self-concept, of Black people, 140, 157

self-recognition: of bias, 66, 67; of privilege, 92–94, 194

sexist language, of Trump, 95

Shakur, Tupac, 4

shaming, 130–31

sharecropping, 202

shoes, 22–23, 244nn20,22

silence about race, as informal rule of racial discourse, 55–62, 79, 81

Simes, Jessica, 59

Sister Space, The, 183–84

sit-ins, 48–49

"slactivism," 177

slavery: effects on White people, 74–75; end of, 204; founding fathers' views on, 56; racist attitudes in justification of, 36; reparations for, 61, 202–3; veiled speech in, 150

Smith, Clint, 56

Smith, Ryan, 63

Snapchat: activism on, 182–83; prevalence of use among adults, 12

SNCC. See Student Nonviolent Coordinating Committee

social anonymity, 43

social desirability bias, 30

socialization, racial, 7, 154

social media: approach to studying racism on, 8–14, 227–37; as both public and private space, 46; and decline of traditional media gatekeepers, 50, 211; as double-edged sword, 47–52; posts about Palestinians on, 208–11, 216; prevalence of use among adults, 12, 176; Trump's ban from, 188–89. See also specific platforms

social media, activism on, 175–85; by celebrities, 214; hashtag, 175–76; participatory politics and, 176–85; personal growth and learning

through, 163–66, 176; rise of, 12–13. See also specific platforms

Solórzano, Daniel, 112

Souls of Black Folk, The (Du Bois), 136, 150

speech. See freedom of speech; hate speech; racist language

Spencer, Octavia, 155

sports: antiracism in, 214; college, 184; fair-weather fans in, 218–19

Staples, Vince, 138–39

Steele, Catherine Knight, 194

Stepansky, Jamie, 228

Stephens-Davidowitz, Seth, 44

stereotypes: of Asian Americans, 207–8; of Black women, 72, 77–78, 163; defined, 32; in implicit bias, 32; performance affected by activation of, 121–22, 141, 152; on Politically Incorrect Confessions, 88–89; Wypipo tweets on, 142–43, 152–53

stereotype threat, 121–22, 141, 152

#StopAAPIHate, 206

#StopAsianHate, 206

street-level bureaucracy, 69

strong Black woman stereotype, 163

structural analysis of racist attitudes, 27, 35–38

structural racism: as antiracist buzzword, 66; evidence of, 103; in internet, 46; use of term on Twitter, 191–92, 193fig.

Student Nonviolent Coordinating Committee (SNCC), 49

subconscious bias, 33–34

subtle (covert) racism, 15–16; challenges of recognizing, 28–29, 106–11; in colorblind racism, 6; in definitions of racism, 27; effects of experiencing,

120–23; reasons for not responding to, 107. *See also* masked racism; microaggressions

Sue, Derald Wing, 107, 115–16, 124, 127

Supreme Court, US, 217

surveys, on racism: admissions of racism in, 20, 23, 25, 30–32; identifying masked racism through, 30–32; implicit bias and, 31–34; modification of questions in, 25, 30–31; school integration in, 37–38; social desirability bias in, 30

systemic racism: challenges of recognizing, 111; discourse in online counterspaces about, 112; use of term on Twitter, 191–92, 193*fig.*, 234–35*fig.*

talented tenth, 150

Taylor, Breonna, 51, 166, 191

tech companies, 46, 50, 216

television, portrayal of racism on, 58–59

TikTok, 12, 216

transformative model of organizing, 170

triangulation, 14

trolls, 42–43

Trump, Donald: admission of vs. acting on racism by, 23–25; anti-Asian comments of, 175, 206; in backlash against Obama presidency, 187; on Charlottesville rally, 24, 143; in conservative politics, 82, 189; on critical race theory, 53; in election of 2016, 23–24, 187–88, 191; in election of 2020, 25–26, 188, 217; immigration policy of, 38, 201; on January 6 riots, 26; overt racism of, 188; policing taskforce under, 56–57; and race-related terms on Twitter, 191, 197–99,

230, 234–35*fig.*, 250n13; racist language of, 9, 16, 95, 175; silence about race under, 56–57; social media bans on, 188–89; Twitter account of, 82, 189; visibility of racism after election of, 187–89; White supremacy and, 16, 24

Tubman, Harriet, 130

Tumblr, 161–63

Tupac, 187

Ture, Kwame, 122

Twain, Mark, 72

Twitter: activism on, rise of, 12–13, 176; anonymity on, 42–43; approach to studying racism on, 12, 14, 227–37; bots on, 42, 241n49; bounty on coding bias in, 216; counterpublics on, 112; fake profiles on, 42–43; on Ferguson uprisings, 50; on Kaepernick, 134; prevalence of use among adults, 12, 13, 176; total number of tweets from 2011–2021, 190–91, 193*fig.*; trolls on, 42–43; Trump's account on, 82, 189. *See also* Black Twitter; Wypipo tweets

Twitter, race-related terms on, 190–211; abolition, 198*fig.*, 199–201, 205*fig.*; All Lives Matter, 204–6, 210*fig.*; Black Lives Matter, 204–6, 210*fig.*; cultural appropriation, 194, 196*fig.*; dog whistle, 197, 198*fig.*; heteropatriarchy, 195, 198*fig.*; ICE abolition, 201, 205*fig.*; intersectionality, 195, 196*fig.*; microaggression, 107, 192, 193*fig.*; model minority, 207–8, 210*fig.*; n-word, 45; Palestine, 208–11, 210*fig.*; patriarchy, 195, 196*fig.*; police abolition/defunding, 200–201, 205*fig.*; racism/racist, 191–92, 193*fig.*, 230, 234–35*fig.*; reparations, 199, 201–3, 205*fig.*; voter

Twitter *(continued)*
    suppression, 197–99, 198*fig.;* White
    privilege, 192–94, 196*fig.*
two-faced racism theory, 34–35, 86–87
Tynes, Brendesha, 51, 82
Tyson, Mike, 213

undocumented immigrants, 38
Unite the Right rally. *See* Charlottesville
    White supremacist rally
universities. *See* college campuses
unmasked racism: defined, 18, 83;
    double-sided consciousness and, 141,
    143; in Politically Incorrect Confes-
    sions, 91, 95–96, 98, 101–2; visibility of
    racism with, 103, 189–90. *See also*
    racism, admissions of; racism,
    expressions of
user-generated content, 45–47
"Uses of Anger, The" (Lorde), 186

Valois Cafeteria, 186
veil, in double consciousness, 139–40, 145,
    149–50
veiled speech, 150–52
Vibrate Higher (Kweli), 209
video games, online: anonymity in, 6;
    Black avatars in, 42; racist language in
    chats in, 1–3, 7
videos: of anti-Black police violence, 48,
    99; of Palestinians, 164–65, 208–9, 216;
    of racism, effects on White people,
    164; of White people calling police on
    Black people, 125
violence. *See* police violence; racial
    violence
virtual reality, 164

visibility, of Whiteness, 142. *See also*
    racism, visibility of
voter suppression: in conservative
    politics, 199, 212, 217–18; use of term
    on Twitter, 197–99, 198*fig.*
voter turnout, 177, 188, 199
Voting Rights Act of 1965, 187

Washington, George, 55–56
Wheel of Time, The (Jordan), 131
White feminists, 195
White fragility, 75–76, 127
White intellectual alibis, 64, 66, 67
Whitelash, 187
White nationalism, 16–17
Whiteness: invisibility of, 142; othering
    of, 142–43
White privilege: defined, 192; in institu-
    tional racism, 122; self-recognition of,
    194; technology used to challenge,
    102; use of term on Twitter, 192–94,
    196*fig.*
White rage, 187
White supremacists: attempts at
    mainstreaming, 17, 238n21; conserva-
    tives' connection to, 212; fake Twitter
    profiles of, 42–43; on fascism, need for,
    212; freedom of speech for, 62–63;
    intersectional approaches to racism
    by, 17; Jim Crow vs. contemporary, 15,
    238n21; racial violence by, 17; as
    Racists, labeling of, 25–26; rallies of,
    24, 62–63, 143, 238n23; spread of
    ideologies online, 6, 45–46; Trump's
    election as win for, 188; Trump's links
    to, 16, 24
Whose Streets? (documentary), 48

Wikipedia, 47

Williams, Apryl, 51, 102

Williams, David, 16

Williams, Jason, 102

Wilson, William Julius, 27, 35

woke, 66, 192

women. *See* Black women

women-centered model of organizing, 169–70

Wood, Roy, Jr., 53–54

World War II, reparations after, 201–2

Wypipo tweets, 142–58; as activism vs. resistance, 157–58; on civil rights movement, 146–47; consciousness raising through, 148–49; externalizing of racism in, 157; meaning and use of term "Wypipo," 142–45; origins of, 143–44; othering of Whiteness through, 142–43; raising awareness of racism in, 152–56; rejection of White superiority in, 152–53; rules of racial discourse broken by, 147–48; solidarity created by, 157; stereotypes in, 142–43, 152–53; trends in number of, 143, 144*fig.;* veiled speech in, 151–52

X, Malcolm, 146, 195

YikYak, 88

Yosso, Tara, 112

"You will not replace us," 16, 238n23

Zembylas, Michalinos, 64

zero-tolerance policies on racist language, 68–70

Founded in 1893,
UNIVERSITY OF CALIFORNIA PRESS
publishes bold, progressive books and journals
on topics in the arts, humanities, social sciences,
and natural sciences—with a focus on social
justice issues—that inspire thought and action
among readers worldwide.

The UC PRESS FOUNDATION
raises funds to uphold the press's vital role
as an independent, nonprofit publisher, and
receives philanthropic support from a wide
range of individuals and institutions—and from
committed readers like you. To learn more, visit
ucpress.edu/supportus.